SEAWEEDS
OF THE WORLD

SEAWEEDS
OF THE WORLD

A GUIDE TO EVERY ORDER

John H. Bothwell

PRINCETON UNIVERSITY PRESS
PRINCETON AND OXFORD

Published in 2023 by Princeton University Press
41 William Street, Princeton, New Jersey 08540
99 Banbury Road, Oxford OX2 6JX
press.princeton.edu

Conceived, designed, and produced by
The Bright Press
an imprint of The Quarto Group
The Old Brewery, 6 Blundell Street,
London N7 9BH, United Kingdom
T (0) 20 7700 6700
www.quarto.com

Library of Congress Control Number: 2022942091
ISBN: 978-0-691-22854-9
Ebook ISBN: 978-0-691-24621-5
British Library Cataloging-in-Publication Data is available

Publisher James Evans
Editorial Director Isheeta Mustafi
Art Director James Lawrence
Managing Editor Jacqui Sayers
Publishing Operations Director Kathy Turtle
Project Editors Anna Southgate, Emily Angus
Design Ginny Zeal
Picture Research Jane Smith
Illustrations John Woodcock

Cover photos, clockwise from top left: SR_002_Mike Guiry/Algaebase; Aikawa Keiko/
Shutterstock; Aikawa Keiko/Shutterstock; SR_002_Mike Guiry/Algaebase; Jiang
Zhongyan/Shutterstock; Olga_Ionina/Shutterstock; HedvikaMichnova/Shutterstock;
Nikox2/Shutterstock; SR_002_Mike Guiry/Algaebase; akepong srichaichana/
Shutterstock; SomprasongWittayanupakorn/Shutterstock; TobyHammond/Shutterstock.

Printed in Singapore

10 9 8 7 6 5 4 3 2 1

CONTENTS

INTRODUCTION

Seaweeds rarely get the affection they deserve. The Roman poet Horace wrote that "Without money, birth and virtue are more worthless than seaweed." A common English name for them is wracks, which comes from the same root as the word "wreck."

But seaweeds are not wrecks, nor worthless, washed-up things. They are astonishingly diverse, vibrant, and important. The soft fronds of jelly weeds are delicate enough to be eaten as noodles, while on the other side of the Pacific Ocean, their kelp relatives can grow as tall as the giant redwoods that they face along the length of the northwest coast of America. The seaweeds provide the foundations for coastal ecosystems from the tropics to the poles, they support sparkling worlds of animal and microbial biodiversity, and they are physically strong enough to absorb wave action and slow coastal erosion along shorelines the world over. They have provided food, fodder, and fertilizer to coastal communities for thousands of years and feed large swathes of island populations to this day.

We should not, however, take seaweeds for granted. The coastal communities that seaweeds support are moving out of balance as climate

change continues its bite and as humans gamely exploit, but inexpertly manage, ecosystems that are millions of years old. Increasing numbers of seaweed blooms are being reported in the Caribbean, the Yellow Sea, and the Atlantic, with beaches becoming ever more clogged as unprecedented quantities of biomass wash up on them. Our best models predict that these seaweed blooms will spread to the North Atlantic, the Mediterranean, and the shores of Africa and South America, all of which are beginning to see an increase in beach-cast weeds. The seaweeds are as integral to our shores as plants and trees are to dry land: if we want to understand the world around us, we need to recognize, respect, and appreciate their diversity.

ABOVE | Fronds of the red seaweed *Asparagopsis taxiformis* rise above the flatter, forked fronds of the brown seaweed *Dictyota dichotoma*.

WHAT ARE SEAWEEDS?

Life on today's Earth is driven by photosynthesis. About half of that photosynthesis is carried out by an evolutionary group called the "land plants," which are our trees and grasses and flowers. The other half is carried out by an older, broader group of organisms called the algae. The algae are astonishingly diverse, but some of them are large and look plantlike and live in the sea, and these "macroalgae" are our seaweeds.

THE PARTS OF SEAWEEDS

The body of a seaweed is called a frond or thallus: an ancient Greek word for a young plant shoot. In many seaweeds, this is divided into three distinct sections: the charmingly named holdfast, the stipe, and the blades. The holdfast tethers a seaweed to the rock or sand beneath it. Holdfasts take many shapes but usually look like large suckers or a tangled collection of roots. Not all seaweeds have holdfasts: some are free-floating,

with the best known being the *Sargassum natans* and *Sargassum fluitans* that grow holdfast-free thousands of miles from any shoreline in the mid-Atlantic's Sargasso Sea.

In larger, tethered seaweeds, the holdfast often develops into a prominent stalk, called the stipe, which connects the holdfast to the fleshier, leaflike body of the seaweed. This fleshy part is called the blade. Different blades may be leafy or straplike, or bushy and branched.

Botanically, a thallus is a structure that is relatively undifferentiated and seaweed thalli have simple internal architectures. Many species consist of only a couple of cell types, with larger species having slightly different outer and inner cell layers that can develop into visible structures such as air bladders. Some cells are able to divide continuously and these form a meristem—the growing core of each seaweed. When seaweeds become fertile, they develop organs that contain

RIGHT | The parts of a seaweed, illustrated using a brown kelp. The inset shows the tangled root type of holdfast (the tangled roots are called haptera, from the Greek for "attachments").

Blade

Stipe

Holdfast

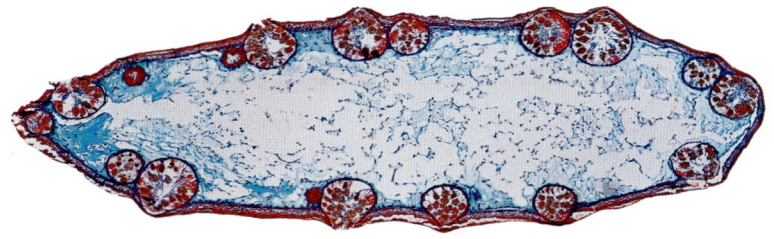

reproductive cells and often appear as blisters on the thallus. Seaweed reproduction is complex so the names given to reproductive structures differ across different species.

By way of comparison, land plants are much more highly differentiated and can contain 50 or more cell types (humans have more than 200), arranged into organs that are connected by extensive vascular systems to allow nutrients from the soil to flow to the leaves and carbon from the air to flow to the roots.

Seaweeds absorb both their nutrients and their carbon from the same place: the water around them. Radioisotope work has shown that the spongelike architecture of some seaweed species does allow a degree of nutrient flow between the dividing meristem and the outer blade. This means that seaweeds have much less need of differentiated organs than plants do and mostly live without them.

EARLY EVOLUTION

We said that the seaweeds were diverse; we will now discover how diverse. The land plants are drawn from one evolutionary group, a phylum called the Embryophyta. The seaweeds, on the other hand, come not from one evolutionary group, but at least three: the Rhodophyta, the Chlorophyta, and the Phaeophyceae. These are more commonly and respectively called the red, green, and brown seaweeds. Each forms either its own taxonomic phylum or, in the case of the Phaeophyceae, a smaller evolutionary group called a class. Why do we group such wildly different organisms together under the same word: seaweed? To answer that question, let us digress to consider how life on Earth has evolved.

THE FIRST SINGLE-CELLED ORGANISMS

The Earth is around five billion years old, and life began a little under four billion years ago. A billion years is an almost incomprehensibly long time: anatomically modern humans are thought to have existed for around the past 200,000 years, which is a mere two hundredths of one percent (0.02 percent) of a billion years. To put that very small number into perspective, the last 12 months were roughly 0.02 percent of the time that has elapsed since the building of the Great Pyramid of Giza: a billion years is a very, very long time.

Nonetheless, five billion years ago, the early Earth solidified from a swirl of cooling stardust. As it cooled, water condensed, clouds formed, and it began to rain. It continued to rain for a billion years, pouring water into the oceans in which life began in the Archean eon, probably around 3.8 GYA (GYA = giga years ago = billion years ago). What form this early life took is much debated, but it would probably have comprised ribonucleic acid (RNA) polymers. Like DNA, RNA is a nucleic acid that can store information in its sequence, but RNA is more reactive than DNA so these early polymers could catalyze simple chemical reactions. Over time, these information-storing catalytic polymers became enclosed in protective membranes, giving the first single-celled organisms.

Where did this early life get its energy from? The simple answer is electricity. Electrical circuits work by gathering negatively charged electrons at one place (the negative terminal) and removing them from another place (the positive terminal). Like charges repel each other, so when the terminals are connected, electrons flow down the connecting wire from the negative terminal to the positive terminal. As the electrons move, their energy can be harnessed and put to useful work.

Living things generate electricity using a very similar trick. Electrons form the outer shells of atoms and when two atoms join together they do so by sharing their outer electrons in a chemical bond. The electrons in some bonds are forced together more closely than in others. The trick that early life discovered was to move electrons from electron-dense chemical bonds into less electron-dense bonds, harnessing the energy as the electrons flowed from one bond to the other.

OPPOSITE | An 1879 illustration from the Spanish periodical *El Mundo Ilustrado*. Nos. 2–4, 6, and 7 are red seaweeds, 8–10 are brown seaweeds, 1 is a green freshwater weed, and 5 is now known to be a bacterial colony.

There are lots of electrons at the negative terminal and these want to get away from each other

Negative terminal

As the electrons flow through the lamp from the negative to the positive terminal, they do work that is converted into light

electron flow

electron flow

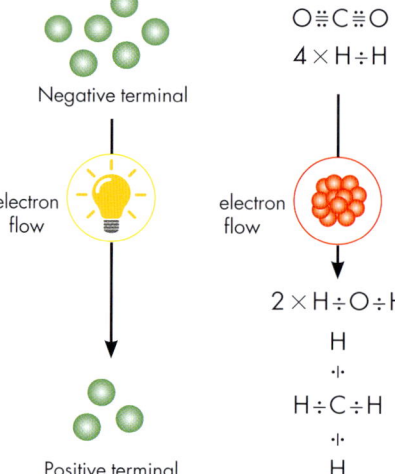

$O \vdots C \vdots O$

$4 \times H \vdots H$

The electrons (shown here as dots) in the C = O bonds are close together and don't want to be (although the oxygen wants to hold onto them)

The cell harnesses the energy of flowing electrons

$2 \times H \div O \div H$

H
$\cdot | \cdot$
$H \div C \div H$
$\cdot | \cdot$
H

The electrons in these chemical bonds are not as close together, so the electrons are happier

RIGHT | Life is powered by the flow of electrons between chemical bonds.

Positive terminal

For about a billion years life powered itself in this electrochemical fashion, prizing apart simple molecules, letting the electrons flow through its cells to power life's metabolism, and then handing the electrons back to new, less electron-dense bonds between the same atoms. For example, one group of early organisms, called methanogens (methane makers), harvested electrons from the bonds in carbon dioxide (CO_2) and hydrogen (H_2) and handed them back to the rearranged carbon, hydrogen, and oxygen atoms to make methane (CH_4) and water (H_2O). Early life proved very good at this simple electrochemistry, evolving into today's bacterial and archaeal lineages, which between them are capable of breaking down around a hundred different simple organic and inorganic chemicals. The microbes that live on sulfates in hot springs or on the hydrogen sulfide released by seafloor hydrothermal vents are both their literal and spiritual ancestors.

THE GREAT OXIDATION EVENT

Although simple molecules were present in the early oceans, they were not abundant and early life

lived on molecular scraps. All the while, it was sitting in a literal sea of unobtainable electrons. Water, H_2O, is an excellent source of electrons because its single oxygen atom forms electron-dense bonds with each of its two hydrogens. These electrons can drive relatively strong electrical currents if they can be prized away from oxygen's clutches.

Unfortunately, cracking open a water molecule to get at its electrons is a difficult trick because oxygen likes to hold on to those electrons. The metabolic breakthrough came eventually, around three billion years ago, with the evolution of photosynthesis by a group of organisms called the cyanobacteria (blue-green bacteria). Over many millions of years, the cyanobacteria evolved complex electron transfer chains: a sequence of reactions that allowed them to use the energy in sunlight as a crowbar. The cyanobacteria crowbarred water open, diverted its electrons through their electron transfer chains to power themselves, and then handed the exhausted electrons back to water's oxygen, creating and releasing oxygen gas. Photosynthesis was such a successful adaptation that cyanobacterial growth took off and began to reform the world, releasing so much oxygen that, slowly but surely, they changed Earth's atmosphere from a methane- and carbon-dioxide-rich one into the one that we have today, in which oxygen makes up around one-fifth of the air around us. This terraforming of the Earth's atmosphere by the cyanobacteria is now known as the Great Oxidation Event and we think it lasted from around 2.5 to 2 GYA, or around half Earth's lifetime ago.

The cyanobacteria did not have it all their own way, however. Living things are engaged in a constant tussle for Earth's limited resources: when one species finds a new trick, others must find a way to adapt if they are not to be outcompeted.

In the case of photosynthesis, the other bacterial species elegantly turned the cyanobacteria's own success against them by using the oxygen (O_2) gas that the cyanobacteria released. We have mentioned that oxygen strongly attracts electrons, which makes the O_2 molecule very reactive because it can pull electrons away from other molecules. This, incidentally, is why you need O_2 to burn things: a flame is simply the heat that results as oxygen very rapidly rips electrons away from the fuel. So, the trick that other microbes developed was to generate energy by burning simple compounds in the cyanobacterial-released O_2 and this metabolic riposte evolved into the process that we now call respiration.

BELOW | The cyanobacterium, *Stigonema*, can form simple branched colonies. Each filament is around 10–20 micrometers wide, so mats of *Stigonema* look like fine gauze.

RIGHT | Cells of the cyanobacterium *Aphanothece*. Each of the individual cells is a just a few micrometers wide and they cluster together to form a simple colony.

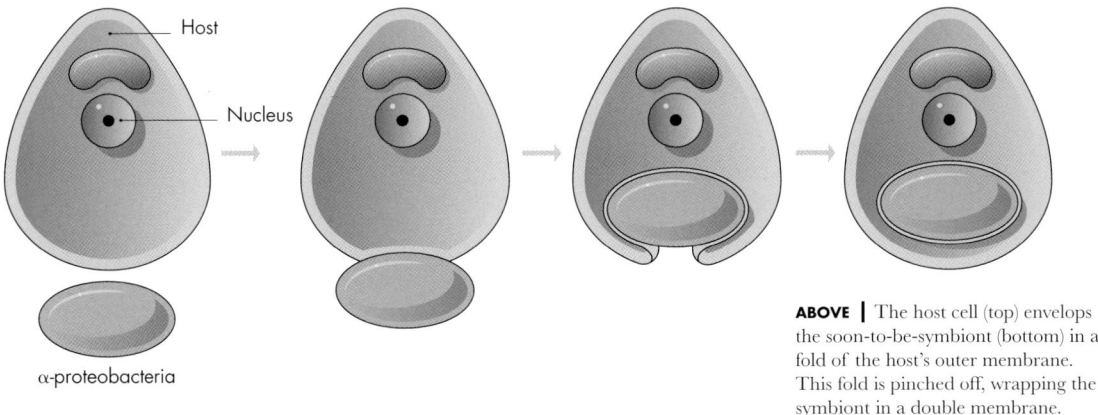

ABOVE | The host cell (top) envelops the soon-to-be-symbiont (bottom) in a fold of the host's outer membrane. This fold is pinched off, wrapping the symbiont in a double membrane.

ENDOSYMBIOSIS

Respiration and photosynthesis are powerful weapons to have in one's armory when competing for limited resources, and other microbes soon wanted in. The exact details remain a matter for debate but, sometime between 1 and 2 GYA, a campaign of cellular piracy took place in which one group of organisms engulfed α-proteobacteria that were capable of respiration. However, instead of being digested, the engulfed α-proteobacteria survived inside their captors and established a symbiosis. Because one species is living inside, rather than alongside, the other, this is called an internal symbiosis or endosymbiosis. This was a "primary" endosymbiosis: one host engulfed something that had previously been free-living.

Slowly and gradually, the new endosymbiont adjusted to its surroundings: it kept both its own cell membrane and the membrane that its host had used to engulf it, leaving it wrapped in a characteristic double membrane. The host and endosymbiont began to synchronize their cells by moving a number of genes from the endosymbiont's DNA across to the host's DNA until, at last, they stopped being two separate organisms and became one single,

compartmentalized cell. Other compartments followed and the host DNA became wrapped in a membrane of its own to form a distinctive cell organelle called the nucleus. The nucleus sitting inside the cell looks, through a light microscope, a little like the stone sitting inside a fruit—so much so that an early-twentieth-century biologist called the possessors of such organelles "true stoned" cells, or "eukaryotes" (Greek: *eu* = "true"; *karyon* = "stone or nut of a fruit"). Today, almost every creature we can see with our naked eye, including ourselves, is a eukaryote and our cells still contain the double-membraned organelles that were once α-proteobacteria. We now call those organelles mitochondria and they continue to respire using the same trick that bacteria developed over two billion years ago to compete with cyanobacteria.

The story of endosymbiosis all sounds extremely unlikely and, indeed, it is extremely unlikely. But a billion years is a very, very long time and, given long enough, unlikely things can and do happen. It is thought that, as with many long-lasting relationships, the establishment of this primary endosymbiosis did not happen overnight: instead, the host and symbiont would have learned first to live in association before the symbiont became attached, and then protected, and then, finally, enveloped.

EUKARYOTIC ALGAE APPEAR

When endosymbiosis works, one lineage gains an entirely new set of abilities. There is no need to painstakingly build something from scratch because one can just steal a whole new metabolic pathway in one fell swoop. It is no surprise, then, that after acquiring mitochondria, some organisms tried the same trick again. Perhaps around 1 or 1.5 GYA, they engulfed and kept hold of cyanobacteria, creating a second organelle that became the chloroplast, in addition to the mitochondria that they already had. Again, this was a primary endosymbiosis in which the host cell engulfed and kept a previously free-living organism and this time it resulted in an organism that could both make food from light and then burn that food to power itself in the dark. These were the first eukaryotic algae: single-celled marine species that were probably a few tens of micrometers across and could make food from light or, in the jargon, were photoautotrophs (light-self-feeders).

These early microalgae (single-celled algae) contained chloroplasts that had been obtained from a primary endosymbiosis and they separated, perhaps between 1 and 1.5 GYA, into what we now know as the red and green lineages: the Rhodophyta and Chlorophyta, respectively (the Phaeophyceae, or brown algae, evolved later, see page 18). Each of the red and green lineages then diversified further into the various red and green orders and genera that we see today. There is no known fossil record of these first microalgae, but they would probably have resembled today's phytoplankton and, like them, would have lived in the top 330 ft (100 m) or so of the ocean's surface.

RIGHT | An electron microscope image through a eukaryotic algal cell, showing the various compartments inside the cell. This cell is from the freshwater species *Chlamydomonas* and is about 10 micrometers wide.

MULTICELLULAR LIFE FORMS

By around 1.1 GYA Earth's oceans were teeming with red and green unicellular eukaryotic life: the microalgal phytoplankton. The continents that had formed perhaps 3 GYA in the late Archean eon had been weathering for almost two billion years and the runoff was accumulating in the oceans to make them salty. It is just after this point, at around 1.03 GYA, that seaweeds first appear in the fossil record. Seaweeds are simply algae that are made up of lots of cells, rather than one cell. They are multicellular, rather than unicellular, and are called the macroalgae to distinguish them from their unicellular microalgal relatives.

Several evolutionary lineages have independently made this jump from one cell to many, for a range of different reasons. Some organisms, most notably bacteria, can switch between unicellular and multicellular assemblages: when bacterial densities grow too high and nutrients become limited, many bacteria will release chemical signals to indicate that they are beginning to starve. If enough of the bacteria in a population are releasing these signals, then the bacteria flip a series of molecular switches and secrete extracellular polymeric substances—put simply, molecular glue. The bacteria aggregate and become bound together by this glue to form bacterial mats, or biofilms. Biofilms made by marine microbial species can grow one on top of the other, rather like a stack of pancakes, to form large and resilient structures called stromatolites. Fossil stromatolites date back to around 2.8 GYA and bacteria still make them today, although they are hard to find: the best-known stromatolites are from Shark Bay in Western Australia, where each stromatolite stack is about the size of a chair.

Unicellular organisms can also form colonies by joining themselves together in a row, one after the other like beads on a string: we call this a uniseriate architecture. Many bacterial species can be uniseriate, and it is likely that the first seaweeds would have had similar architectures. Such simple filaments are the starting points for more complex structures, all of which we see in modern seaweeds. Filamentous genera, such as the brown *Pylaiella*, form tangles that can be unbranched or branched. Filaments may lose their internal walls and become woven together to form

A POSSIBLE PATH TO MULTICELLULARITY

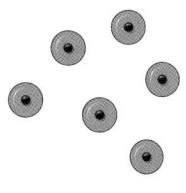

Single-celled organisms

From unicellular to 1-D multicellular

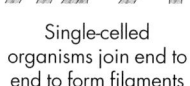

Single-celled organisms join end to end to form filaments

1-D multicellular to 2- and 3-D multicellular

Filaments weave together to form more complex structures

Bangiomorpha pubescens

Proterocladus antiquus

wickerwork-like structures, as in the green *Codium*, or they may divide and spread in two directions rather than one, to give flat sheets, as in the green *Monostroma*, or in three directions, to give the fully three-dimensional kelps.

It is currently thought that simple multicellular filamentous algae, that is true seaweeds, first appeared independently in both the red and green lineages a little over 1 GYA. At the time of writing, the earliest taxonomically resolved multicellular eukaryotic fossil is of a now-extinct red seaweed species called *Bangiomorpha pubescens*, which is around 1.03 GYA. *Bangiomorpha* shows a filamentous branched structure with some differentiation at the tip of the filament into what is presumably a collection of reproductive cells. It is strikingly similar to modern red species such as *Polysiphonia*. The oldest currently known green seaweed is slightly younger, at around 1 GYA, and is of another now-extinct species, *Proterocladus antiquus*. This looked a little like today's green *Cladophora* species, only much smaller at around 2 mm long. The fossil record from a billion years ago is understandably extremely patchy, so these timelines will likely change as further discoveries are made.

ABOVE | The 1.03-billion-year-old red seaweed *Bangiomorpha pubescens* is now extinct, but bears a similarity to present-day *Polysiphonia*. Also extinct, the 1-billion-year-old green seaweed *Proterocladus antiquus* bears a similarity to *Cladophora*.

BELOW | Cyanobacterial stromatolites in Shark Bay, Western Australia.

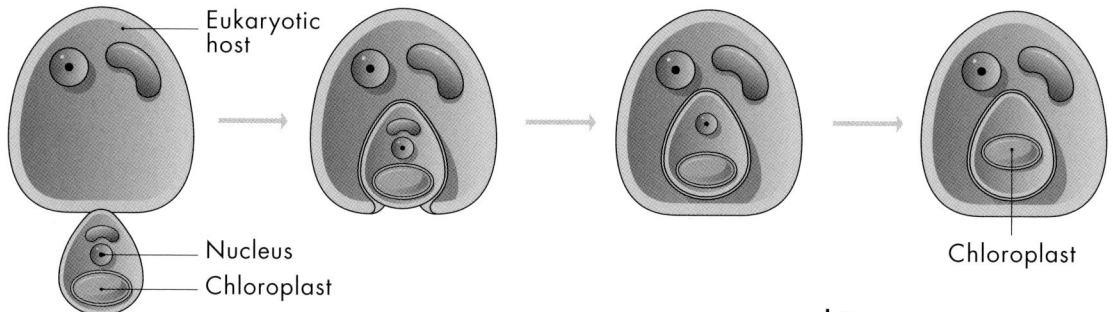

Eukaryotic host

Nucleus

Chloroplast

Eukaryotic proto-symbiont

Chloroplast

ABOVE | The host eukaryote envelops the soon-to-be-symbiotic eukaryote. Most of the enveloped eukaryote's organelles then degrade, leaving the desired organelle (in this case, the chloroplast) wrapped in four membranes. For simplicity, the mitochondria are not shown.

A SECONDARY ENDOSYMBIOSIS

The third lineage of the seaweeds, the Phaeophyceae (brown seaweeds), are only around half the age of the reds and greens, first appearing perhaps 600 MYA (million years ago). They are unusual for two reasons: first, they evolved from a completely separate eukaryotic lineage to the reds and greens and second, they gained their chloroplasts not from a primary, but from one or more "secondary" endosymbioses. That is to say, the proto-brown host engulfed something that had itself engulfed something previously free-living.

The proto-brown host came from a group that evolved into what is commonly called the supergroup Stramenopila. Eukaryotic life is usually classified into around half a dozen major groups. It is generally agreed that animals and fungi sit in one of these major groups, called the Opisthokonts. The brown algal-containing Stramenopila is another of these major groups and the green and red algae sit in a third, called the Plantae (along with the land plants). The rest of the major groups, the Rhizaria, the Amoebozoa, the Excavata, contain various common, but microscopic, amoeba-like

organisms. The Stramenopila, therefore, were originally about as different from the red and green algae as humans are.

The proto endosymbionts, on the other hand, are thought to have been red algae. The exact details of the process are the subject of current research, but it is thought that the proto-brown host engulfed one or more red algae around 600 MYA and retained the chloroplasts, giving it the ability to photosynthesize and setting it on the path to a sedentary photoautotrophic lifestyle.

While primary endosymbiosis leaves its resulting organelles wrapped in two membranes, secondary endosymbiosis leaves its resulting organelles wrapped in four. The two inner ones are those that surrounded the original red algal chloroplast, a third is derived from the red alga's cell membrane (the rest of the alga is discarded because the important bit is the chloroplast), and the final, outer, membrane is the remains of the brown host's engulfing cell membrane.

INVASION OF THE LAND

As always, exact dates are hard to estimate with any confidence, but around 900 MYA, the green algae separated into two groups that are today called the chlorophytes and the streptophytes. Branches in both groups turned from plankton into weeds at various points over the next few hundred million years. However, an important distinction had arisen by the Cryogenian-Ediacaran periods (around 720–540 MYA): while most chlorophyte algae remained marine, the streptophyte algae were beginning to adapt to the waters found inshore near river estuaries.

By this time, billions of years of weathering had turned the river estuaries of the Neoproterozoic (around 1,000–541 MYA) into rich sources of the minerals and elements needed for plant growth. Slowly, but inexorably, one branch of the streptophyte seaweeds moved closer and closer inshore. The more they became adapted to freshwater, the farther they were able to penetrate up rivers. They ran the risk, of course, of being washed up on the banks of those rivers but they began to evolve waxy, protective coats that would protect them from occasional desiccation. In fact, those individuals that were washed up found that the wetlands and riverbanks were extremely nutrient rich and natural selection began to favor those who could hold themselves in place with simple rootlike structures. They began, in short, to evolve into the land plants, making their earliest moves toward the early Ordovician period, perhaps 470 MYA, and starting the plant colonization of the land that built over the next 50 million years or so, through the Silurian period (440–420 MYA) to the early Devonian (around 415 MYA). So, not only were seaweeds the first multicellular organisms to persist on our planet, they also gave rise to the land plants that now support humanity.

TAXONOMIC VARIATION

In summary, there are three major groups of seaweeds: the reds, the greens, and the browns. The reds and greens evolved as single-celled organisms around 2 GYA. Some members of the greens and reds stayed unicellular and are now phytoplankton, but some branches of both the red and green algae became multicellular and are now the red and green seaweeds. More recently, and perhaps 600 MYA, eukaryotic Stramenopiles engulfed unicellular reds to give rise to the brown algal lineage. As with the greens and reds, some members of the browns remained unicellular and are now phytoplankton (e.g. the diatoms), while other members became multicellular to give rise to the brown seaweeds. In evolutionary jargon, the seaweeds are joined together horizontally, by endosymbiotic gene transfer from one lineage to another, rather than vertically, which would have involved gene transfer from parents to offspring.

Despite being from very distinct evolutionary lineages, all seaweeds have chloroplasts, so all have adopted the same essentially sessile photosynthetic lifestyle and body shapes, but it has either been billions of years since they were the same (reds vs greens) or they were very different to begin with (browns vs greens and reds) and those differences persist today in the fundamental structure and chemistry of each seaweed group.

For that reason, the red, green, and brown seaweeds have major differences in their

BELOW | A salad of green, red, and brown seaweeds, washed up on a pebbly shore.

biochemical and ultrastructural building blocks. Biochemically, the different lineages have evolved different pigments, which explains their different colors and gives each group its name. These pigments help to absorb light for photosynthesis (see page 38). The different seaweed lineages are also physically constructed from different cell wall polymers. All three of the main seaweed groups have cell walls made of cellulose, which they inherited from their cyanobacterial endosymbionts, but the cellulose fibers are glued together by different branched polymers in each of the three groups. In the green algae, some of these are called ulvans; in the reds, some are the carrageenans and agars; and in the browns, some are the fucoidans and alginates. As we will see, the different chemistry of these polymers means that

different groups of seaweeds are suited to different uses. Finally, the red, green, and brown seaweeds have different ultrastructures—that is, the internal structure of the cell. We will not go into too much detail about these ultrastructural differences, although we have already met one: red and green algae have two membranes around their chloroplasts, because those chloroplasts were derived from a primary endosymbiosis, while brown algae have four membranes around theirs, because brown chloroplasts were derived from a secondary endosymbiosis.

BELOW | The algal branches of the tree of life. The seaweeds are not as closely related to each other as we sometimes think.

ALGAL TREE

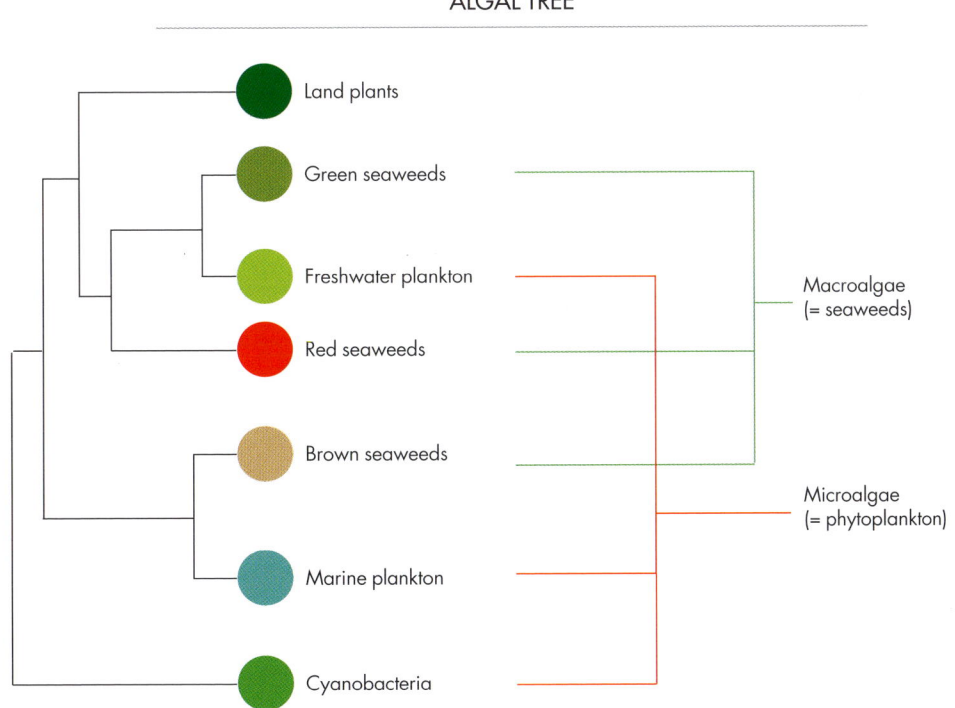

Land plants

Green seaweeds

Freshwater plankton

Red seaweeds

Brown seaweeds

Marine plankton

Cyanobacteria

Macroalgae
(= seaweeds)

Microalgae
(= phytoplankton)

Before looking at how the evolutionary diversity of seaweeds is reflected in their different adaptations and distributions and in the ways in which they contribute to the health of coastal ecosystems, we should pause and remember that seaweeds have interested millions of people for thousands of years in hundreds of countries. What we know about them reflects the interest of amateur naturalists, enterprising merchants, professional scientists, and many, many other people.

Seaweeds have great cultural significance: our understanding of them is not just a way to exploit their resources, but also a way to connect to our pasts and a remembrance of our links to the seas that surround us. An exemplar here is the work of the astonishing Isabella Aiona Abbott, the first native Hawaiian to earn a science doctorate and Stanford's first female biology professor. A preeminent taxonomist of the seaweeds of Hawaii and the Pacific, Abbott also understood the deep links between a coastal people and their seaweeds, describing the sharing of *Limu kala* (*Sargassum echinocarpum*) as a peace rite in much the same way that people in the United Kingdom talk about "breaking bread." As she wrote in her short but influential 1974 book, *Limu: an ethnobotanical study of some edible Hawaiian seaweeds*, "Before the coming of the white missionaries and the intervention of Queen Kaahumanu (1819), certain foods were *kapu* (forbidden) to women, including pork, bananas, coconuts, and a variety of fish […] Hence, rewards came to those who mastered the arts of search and preparation of the non-*kapu* food, which, in this case, were seafoods such as invertebrates and algae."

A NATURAL HISTORY

EARLY PHYCOLOGY

A phycologist is somebody who studies algae. The name comes from the Greek word for seaweed, *phycos*, and the beginning of modern phycology may be dated, somewhat arbitrarily, to 1753. The Enlightenment was at its peak and the torch that had been lit in the seventeenth century was being carried through the eighteenth century by new generations of scholars across every major European country. One of the greatest of those scholars was Carl Linnaeus, who is widely recognized as the founder of modern taxonomy. Linnaeus devised a system of classification that sorted living things into a hierarchy of groups, based on shared characteristics such as how many limbs an organism had.

Linnaeus' system worked well for animals, but less well for plants, in which growth is much more variable. Almost all humans are born with the same number of limbs, for example, but fewer trees have exactly the same number of branches. In 1753, therefore, Linnaeus published his *Species Plantarum*, in which he adapted his classification for use in plants by looking at their reproductive organs, the flowers and seeds, which were more consistently constructed. Unfortunately, he ran into trouble with the seaweeds because they didn't have any recognizable reproductive organs. Linnaeus therefore placed them into his Cryptogams ("hidden seed" plants) along with mosses and fungi and divided them into

three groups: *Ulva*, *Fucus*, and *Conferva*. Into *Ulva* went the leafy species, including those that we now know as the green sea lettuce *Ulva lactuca* and the red *Porphyra umbilicalis*. Into *Fucus*, Linnaeus placed the larger, more robust seaweeds, with his *Fucus* including species that we now know as the red sea oak *Phycodrys rubens*, the brown kelp *Saccharina latissima*, and, of course, the eponymous *Fucus* wracks. *Conferva*, finally, held the smaller, shrubbier species, such as *Cladophora* and *Griffithsia*.

Linnaeus' system was so obviously fundamental that it led to an explosion of interest in the classification of all forms of life. His books went through multiple revisions in Linnaeus' lifetime alone, expanding from a few hundred species in the first edition of his greatest work, the *Systema Naturae*, to more than 4,000 in the twelfth. Linnaeus died in 1798 but by then taxonomy had become too big for any one person to master. Instead, other people picked up its different branches and ran with them.

EARLY MICROSCOPIC STUDIES

Linnaeus' sex-based classification of plants forced researchers to look more closely at how seaweeds reproduced. In his book *Nereis Britannica*, written around the turn of the nineteenth century, the British naturalist John Stackhouse explained: "Being convinced of the insufficiency of the common pillar microscope […] I determined to

OPPOSITE | A nineteenth-century seaweed collector's book of pressed seaweeds. This one was made on Jersey, one of the islands in the English Channel.

LEFT | The Swedish botanist Carl Linnaeus in 1739. He is only in his early 30s here but this is his wedding portrait, hence his bewigged finery.

25

Gigartineae.

Chondrus crispus Lyngb. Gigartina mamillosa J. Ag.

LEFT AND ABOVE | *Chondrus crispus*, shown here in a nineteenth-century German herbal; John Stackhouse (1742–1819), who named it.

furnish myself with higher magnifying powers." This exemplifies how scientific progress was made throughout the nineteenth and early twentieth centuries—through increasingly elegant and detailed microscopic studies of seaweed reproduction.

Stackhouse named or reassigned various species along the English Channel, but his observations and species reallocations were prescient because similar reassignments of Linnaeus' seaweed species were also being started by botanists across Europe. Early-nineteenth-century microscopes and chemical analytical methods were poor by today's standards but were steadily improving and, in France, Jean Vincent Lamouroux, Professor of Natural History at

Caen, was the first to argue that red and brown seaweeds should be separated into their own distinct phyla. His suggestion was adopted and followed by a swathe of taxonomic revisions in the mid-nineteenth century, based on ever-increasing and improving microscopic observations.

By 1845, microscopes had improved to the extent that Gustave Thuret, a gentleman botanist who lived in Paris, could write: "During the excursion which I made with M. Decaisne on the coasts of the British Channel, we had frequent opportunity of studying the spores of *Ectocarpus siliculosus*, of *Ulva lactuca*, and of *Enteromorpha compressa*." He continued: "In the *Ectocarpus* we found two cilia inserted on a colorless rostrum. In *Ulva* and *Enteromorpha* the spores have four

Dictyota dichotoma
in the young state, x

cilia." *Ectocarpus* is a brown seaweed closely related to *Pylaiella* and *Enteromorpha compressa* is the old name for what is now *Ulva compressa*. What Thuret had found and reported was that the brown seaweeds produce reproductive cells that have two hairs, while the green *Ulva* seaweeds produce reproductive cells with four hairs. Thuret called these hairs "cilia," but today they are referred to as "flagella."

The significance of Thuret's work lies in the fact that researchers were realizing that cell structures could be used to classify organisms. This makes a great deal of sense: plants and algae show extremely diverse growth and their visible forms can show a lot of variation. The cells of a species, however, are more consistent building blocks. If the cells of two species look similar, those species are likely to be closely related, even if the two species themselves grow into quite different shapes. Conversely, if two seaweeds look superficially similar, but have dissimilar cells, then they are likely to be only very distant relatives.

ELECTRON MICROSCOPY

Much of our understanding of seaweeds draws on remarkable light microscopy work by many researchers in the latter half of the nineteenth and early part of the twentieth centuries. But light microscopy struggles to resolve objects that sit closer together than about one micrometer, which is about the size of many seaweed reproductive cells. This improved with the twentieth century's invention of the electron microscope. Electron microscopes have a resolution limit around a thousand times better than light microscopes, allowing us to study cell components that are as small as individual protein molecules. Their main drawback, incidentally, is that electron microscopes can only work with dead things, while light microscopes can look at living ones.

Working in Leeds, in the United Kingdom, in the late 1940s, Irene Manton was one of the first scientists to use electron microscopy to probe cell biology. In 1949 she wrote about the brown seaweed, *Fucus serratus*: "The hairy appendages on the front flagellum are completely invisible except with the electron microscope, and they appear to constitute a definite morphological feature by which, in this species, the front flagellum differs from the hinder one."

Manton's work reinvigorated the use of ultrastructural characteristics to study algal biology and classification. For example, if a seaweed produced gametes with no flagella, it must be one of the red Rhodophyta; if it produced gametes with two lateral flagella (coming out of the side), one of which was smooth and the other covered in tiny hairs called mastigonemes, then it was a member of the brown Phaeophyceae; and if it produced gametes with smooth apical flagella (coming out of the end of the gamete), it was a member of the green Chlorophyta. Other ultrastructural features gave insights into seaweed subgroups and this taxonomic and functional work continues to this day.

BELOW | The top of a brown seaweed conceptacle, showing the microscopic hairs that will pull apart to reveal the opening that releases the reproductive cells. This is a false-color scanning electron microscope image.

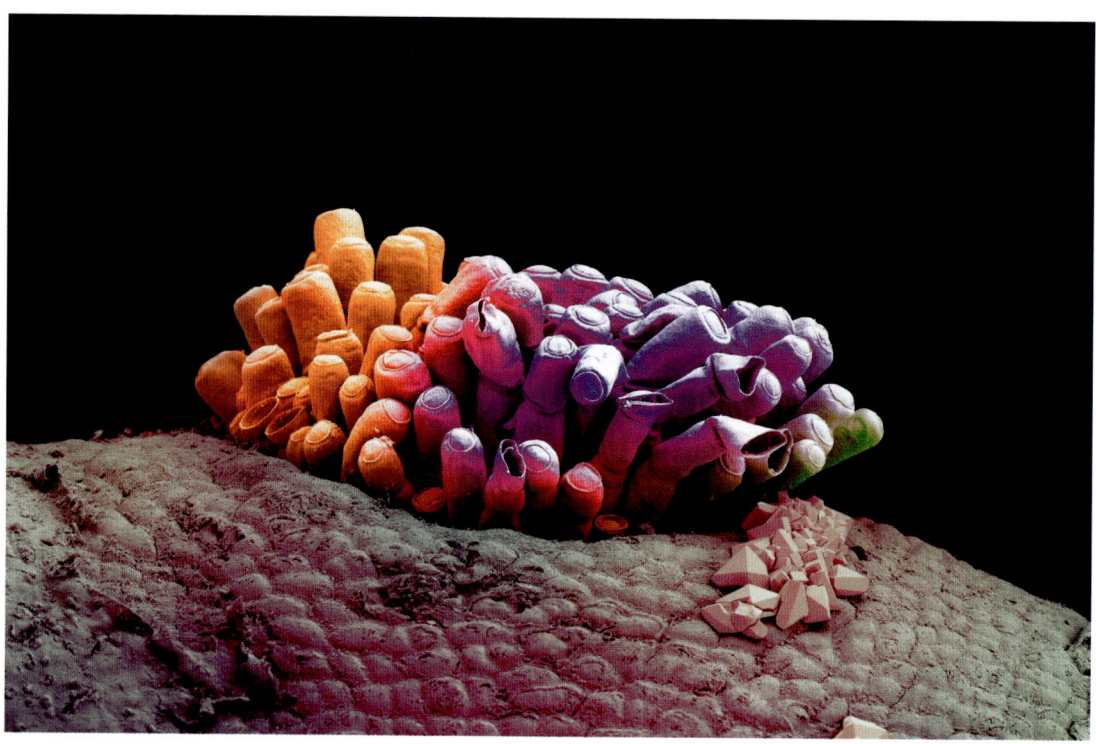

THE MOLECULAR ERA

The twentieth century saw a lot of work on the ultrastructure and biochemistry of seaweeds but, for various reasons, phycology was slower than other disciplines to benefit from the molecular biology revolution that started in the 1940s. Accordingly, much remains unknown about exactly how seaweed genes determine their development and physiology.

Things are changing, however. The past decade has seen the sequencing of around a dozen seaweed genomes, including a red *Pyropia*, a green *Ulva*, and a brown *Saccharina*, and more are on the way. Genome sequencing generates a huge amount of information about an organism but that information is difficult to interpret. As an analogy, every organism is built to a recipe and recipes have two parts: the list of ingredients and the instructions for how to cook those ingredients. We are currently able to read the list of genomic ingredients, but we struggle to understand the instructions for how to use them. The sequenced seaweed genomes each contain around 10,000 genes and range in size from the small (around 100 million base pairs in *Ulva*) to the large (more than 1.5 billion base pairs in some kelps). For comparison, human genomes are around three billion base pairs long and contain around 20,000 genes.

At the time of writing, scientists are still developing molecular toolkits to study seaweed genomes, using approaches such as genome editing and the fluorescent labeling of genes of interest. Our molecular journey into seaweeds is entering an exciting time and there is a lot to look forward to.

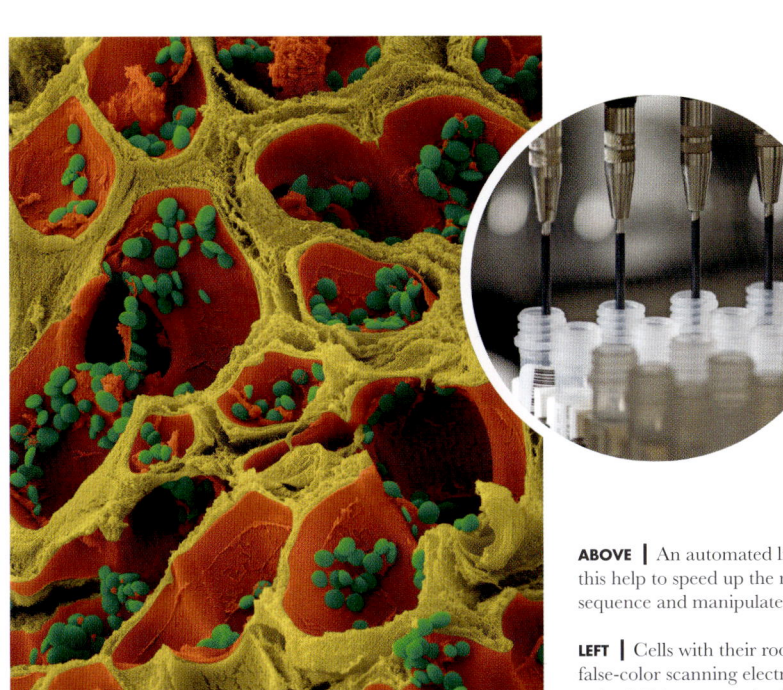

ABOVE | An automated liquid handler. Robots like this help to speed up the rate at which we can sequence and manipulate the DNA of seaweeds.

LEFT | Cells with their roofs ripped off. This is a false-color scanning electron microscope image of red calcifying seaweed cells. Chalk-strengthened cell walls are shown in yellow and chloroplasts are shown in green. Each cell is around 10 micrometers across.

SEAWEED REPRODUCTION

Gustave Thuret's work on the flagella of brown and green seaweeds marks a high point in nineteenth-century phycology, and one on which Thuret himself built. Over the next couple of decades, he and others systematically characterized the sexual systems and life cycles of seaweeds and the picture that emerged was complex. This should come as no great surprise. Remember: seaweeds represent several different phyla spread across two major evolutionary groups, so we would expect them to display a wide variety of life cycles and reproductive habits. This complexity has resulted in much jargon. In an attempt to make seaweed reproduction a little easier to understand, it is useful to outline a number of general principles first.

Organisms consist of generations, which produce reproductive cells, which in turn produce generations, and so on ad infinitum. Generations are longer-lived multicellular forms that exist for the life of the organism, whereas reproductive cells are unicellular and much shorter-lived: their function is simply to give rise to a new multicellular generation. The recipe for making any organism is contained in its genome. Generations and reproductive cells may contain one version of the genome, in which case they are called haploid (single-copied), or two versions

GENERIC SEAWEED LIFE CYCLE

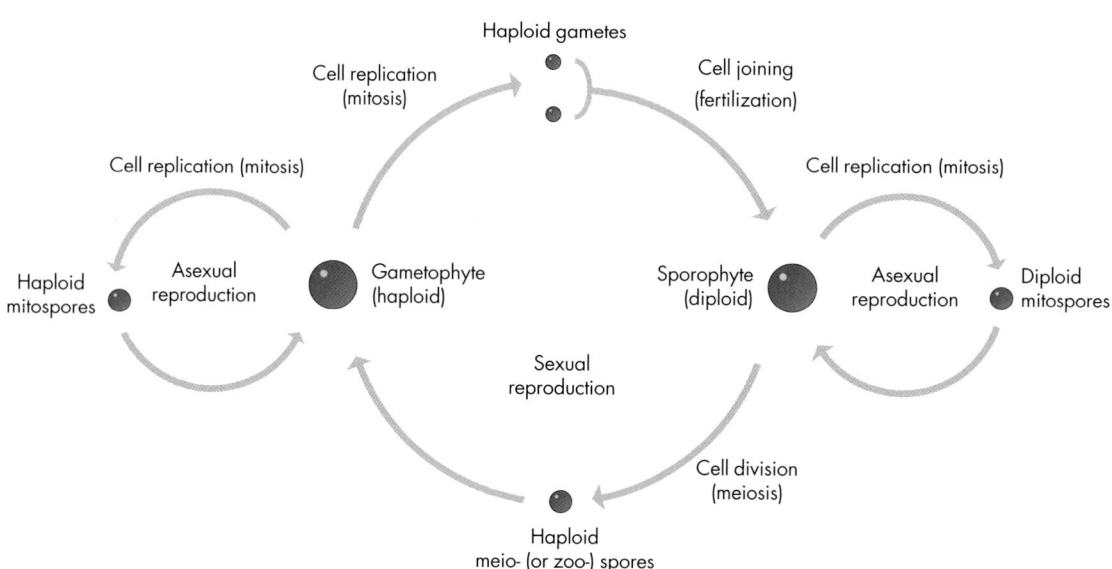

ABOVE | This generic seaweed life cycle alternates between two multicellular generations, a haploid gametophyte and a diploid sporophyte.

(diploid), or three versions (triploid), or even more (polyploid). Reproduction may be asexual, in which a reproductive cell called a spore is able to produce a new generation on its own, or sexual, in which two reproductive cells called gametes combine to give the new generation.

There are two types of gametes, which we usually—but not always—call male and female. Gametes may look the same (isogamous), a little different (anisogamous), or look so different that they are clearly sperm from males and eggs from females (oögamous). Finally, the two types of gametes may be made either by a single individual (monoecy, from the Greek for "single house") or by two separate individuals with distinct male and female sexes (dioecy, from the Greek for "two houses"). In dioecious species, the two sexes may look the same or they may look different, in which

case we say they show sexual dimorphism. There are exceptions to all of this, especially in land plants, but those are the basic concepts.

To practice, let's apply those concepts to the reproduction of the species with which we're most familiar: ourselves. Humans reproduce sexually. Broadly speaking, our multicellular diploid generation comes in one of two sexes—male and female—and each sex produces single-celled haploid gametes that fuse to give a single-celled diploid zygote that divides to grow into a new multicellular diploid generation. We have only one free-living multicellular diploid generation, which is our two-arms-two-legs-one-head shape. We display some sexual dimorphism, but not that much when compared to many animal species. We also have only one class of reproductive cell: haploid gametes. Our gametes come in two flavors, sperm and eggs, so we're oögamous, but that's a minor detail: the more important thing is that humans can only reproduce through haploid gametes.

SEAWEED LIFE CYCLES

The life cycles of seaweeds are much more diverse, and are based around the alternation of two multicellular generations: one haploid generation (usually called the gametophyte) and one diploid generation (usually called the sporophyte). The haploid gametophyte generation may produce two kinds of unicellular reproductive cells, both of which are produced by cell replication (mitosis) and so will also be haploid. The first of these are spores, which form in structures called sporangia and which go on to regenerate the haploid generation through asexual reproduction. The second of these are gametes, which form in structures called gametangia and allow sexual reproduction by fusing with another gamete

(fertilization) to give rise to the diploid sporophyte generation. The diploid sporophyte generation forms unicellular haploid spores in sporangia through reductive cell division (meiosis) and these haploid spores will grow into the haploid gametophyte generation. The haploid and diploid generations may look the same (isomorphic alternation) or may look very different (heteromorphic alternation).

BELOW | The heteromorphic generations of *Cutleria*. The leafy haploid gametophyte is on the left; the crustose diploid sporophyte is on the right.

TAXONOMIC CONUNDRUMS

Heteromorphic alternation of generations has historically confused many taxonomists and there are instances in which a newly identified species was eventually realized to be simply another generation of a previously known species. For example, in 1843 the Italian phycologist Giovanni Zanardini gave the name *Aglaozonia* to a new genus of brown seaweeds. In 1878, the German phycologist Paul Falkenberg demonstrated that *Aglaozonia* was, in fact, the diploid sporophyte generation of an already named, and very different-looking, brown genus called *Cutleria* (see page 176). By taxonomic convention, the name *Cutleria* had been described first, so took precedence. Both forms are now recognized as *Cutleria* but, in a touching nod to its later name, the form that had been called *Aglaozonia* is still called the Aglaozonia phase.

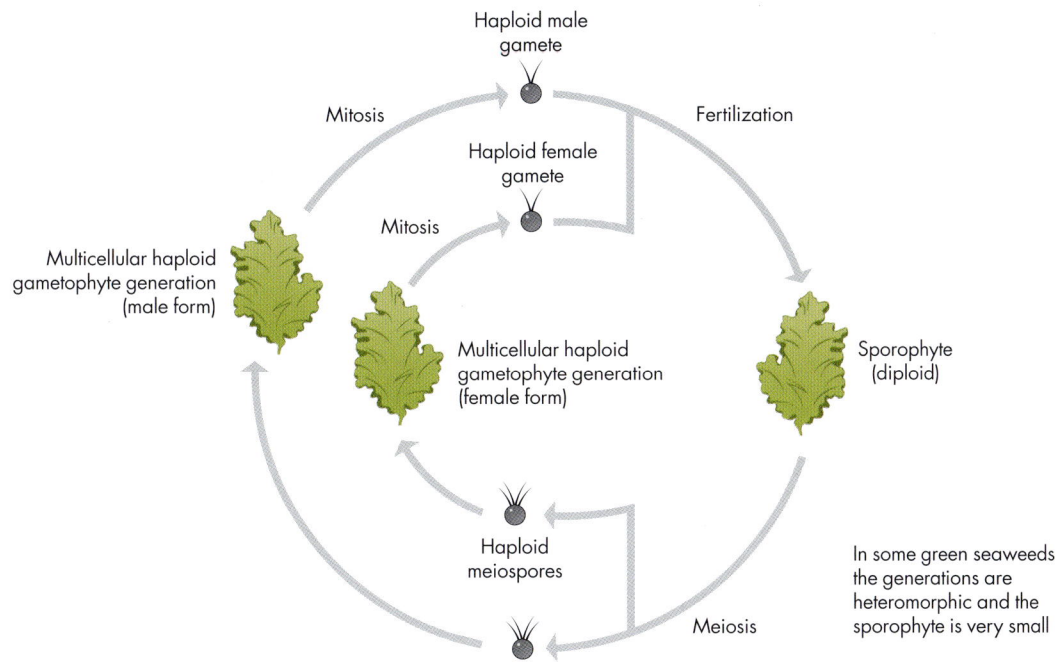

Haploid male
gamete

Mitosis

Fertilization

Haploid female
gamete

Mitosis

Multicellular haploid
gametophyte generation
(male form)

Multicellular haploid
gametophyte generation
(female form)

Sporophyte
(diploid)

Haploid
meiospores

In some green seaweeds
the generations are
heteromorphic and the
sporophyte is very small

Meiosis

GREENS

Some green seaweeds, such as *Codium*, display very simple sexual life cycles like ours. Others can display either isomorphic (two generations, both looking the same) or heteromorphic (with the haploid gametophyte being larger) alternation of generations. In *Ulva*, for example, the generations are isomorphic, while in *Spongomorpha* the seaweed is the haploid gametophyte and the diploid sporophyte is microscopic. *Ulva* sticks closely to the model life cycle: the fertile gametophyte develops gametangia, within which haploid biflagellate gametes form and are released. The flagella on these gametes allow them to move and they swim around to find another gamete with which they can fuse to form the diploid zygote from which the multicellular diploid sporophyte will grow. When the sporophyte becomes fertile in its turn, sporangia form on the thallus and haploid spores

ABOVE | The life cycle of the green seaweed *Ulva* is isomorphic, dioecious, and isogamous.

are generated by meiosis. These haploid spores are the reproductive cells that Gustave Thuret reported in 1845, with four flagella that allow them to swim in much the same way the gametes do. Under the microscope, these spores can look pretty lively and were consequently called zoospores, a term used to refer to any spore that can swim and move on its own; the term meiospore is more common today. These haploid meiospores then settle and regenerate the haploid multicellular gametophyte.

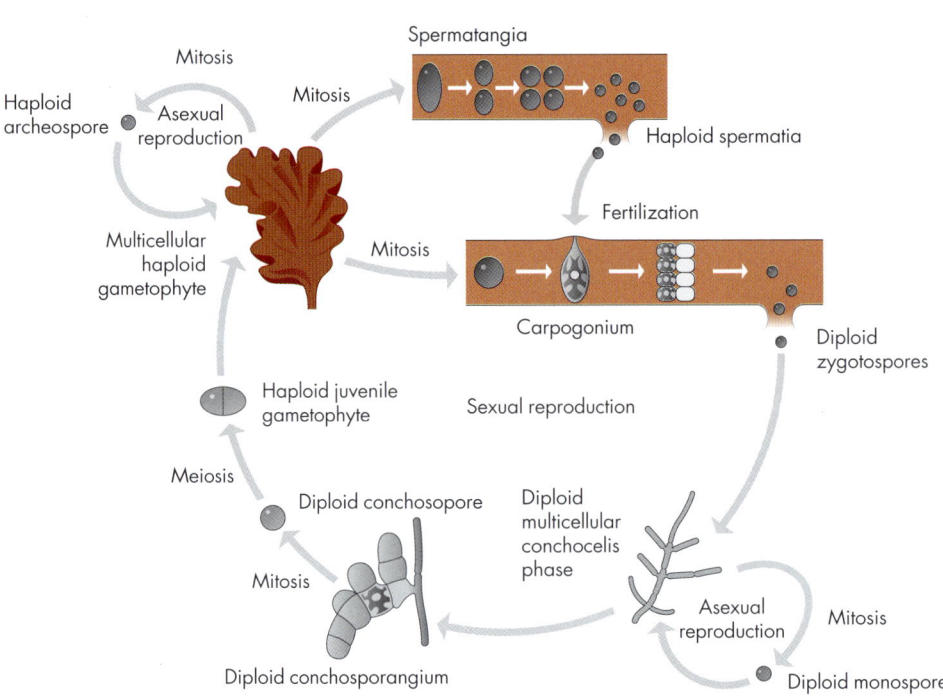

REDS

Red seaweeds are, in part, defined by one very unusual characteristic, which is that none of their cells have any flagella. This means that their reproductive cells cannot move very fast. To signify that, the male cells are called spermatia, rather than sperm. Perhaps to get around this movement problem, although according to some theories perhaps not, the red seaweeds have evolved byzantine sexual life cycles. These vary between different red algal groups but usually conform to one of two basic models: a) heteromorphic alternation of generations or b) modified alternation of generations in which a third multicellular phase is inserted between the gametophyte and sporophyte: this is often called a triphasic life cycle.

Heteromorphic alternation is seen in the red bangiophyte branch (seaweeds such as *Neopyropia*) and starts with a haploid gametophyte. These

develop male structures called spermatangia, which release spermatia, and female structures called carpogonia. A spermatium will fertilize a carpogonium, and the resulting diploid zygote divides inside the gametophyte blade into lots of unicellular diploid zygotospores. Exactly what these zygotospores did remained a mystery until the extraordinary work of Kathleen Drew-Baker in the mid-twentieth century. According to Drew-Baker: "When such filaments, either at germination or at a much later stage, are brought into contact with old shells, they penetrate the shells and develop rapidly into extensive and highly characteristic growths, apparently identical with an alga known for many years as *Conchocelis rosea Batters* and described so admirably by Batters." This diploid conchocelis phase that nineteenth-century phycologists had thought a different species is actually the second of *Porphyra*'s two multicellular generations (similar to the

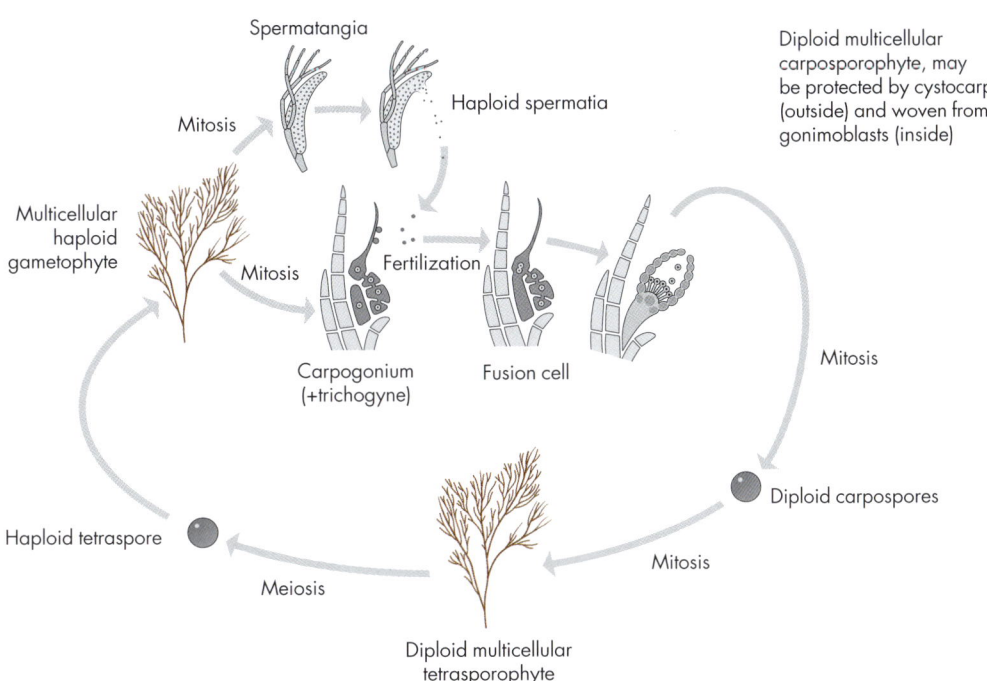

Spermatangia

Mitosis

Haploid spermatia

Diploid multicellular carposporophyte, may be protected by cystocarp (outside) and woven from gonimoblasts (inside)

Multicellular haploid gametophyte

Mitosis

Fertilization

Carpogonium (+trichogyne)

Fusion cell

Mitosis

Diploid carpospores

Haploid tetraspore

Meiosis

Mitosis

Diploid multicellular tetrasporophyte

diploid sporophyte in the greens) and can either reproduce indefinitely and asexually through diploid monospores, or continue along the sexual cycle by developing a structure called the conchosporangium. Conchosporangia produce diploid conchospores that float away, settle, and regenerate haploid gametophytes.

In the other major red seaweed branch, the Florideophyceae, the fertilized carpogonium behaves differently. The carpogonium is cylindrical and tapers into a thin structure called a trichogyne. The trichogyne acts like flypaper and catches spermatia. After fertilization, the diploid zygote does not immediately divide into zygotospores. Instead, it develops into a distinctive additional multicellular generation, called the carposporophyte. Exactly how it does this varies between species, but in general the zygote will fuse with a nearby cell, called an auxiliary cell, to create a multinucleate fusion cell. The fusion

OPPOSITE | The life cycle of the red Bangiophyte genus *Porphyra* is heteromorphic, monoecious, and anisogamous.

ABOVE | The red Florideophyte life cycle is similar to the Bangiophyte's. The fertilized carpogonium often develops into a third multicellular generation, the carposporophyte.

cell divides into diploid filaments called gonimoblasts, which weave together to make the carposporophyte. The tips of the diploid gonimoblast filaments develop into diploid carpospores, which are released and grow into multicellular diploid tetrasporophytes. These produce haploid tetraspores by meiosis, which regenerate the haploid gametophyte.

So, some red seaweeds have two multicellular generations (the haploid gametophyte and diploid conchocelis) while many of the Florideophyceae have three (the haploid gametophyte, diploid carposporophyte, and diploid tetrasporophyte).

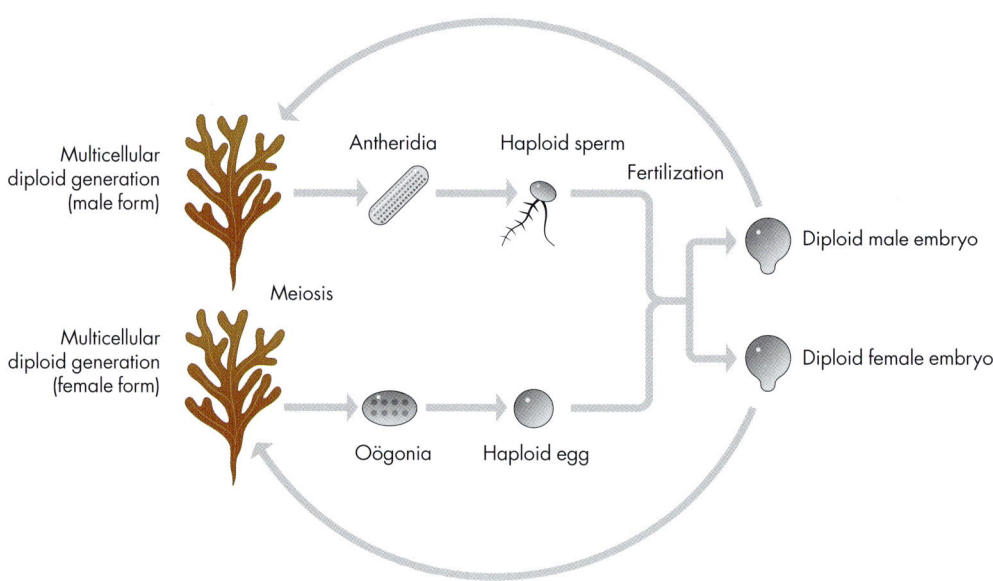

BROWNS

Brown seaweed life cycles are relatively simple. They either have only one diploid generation, as in *Fucus*, or they show isomorphic alternation of haploid and diploid generations, as in *Pylaiella*, or they show heteromorphic alternation with the diploid sporophyte generation being much larger than the haploid gametophyte, as in *Cutleria* or *Macrocystis*.

The fucoids have the simplest life cycle of any seaweed. Their life cycles are almost exactly like ours: the visible seaweed is the diploid generation and, when fertile, these develop specialized visible blister-like structures called receptacles. Receptacles contain smaller structures called conceptacles that develop into either antheridia (gametangia that produce sperm) or oögonia (gametangia that produce eggs). The antheridia and oögonia are released into the water column, where they break open to liberate sperm and eggs.

The eggs release a sperm attractant and the sperm detect this and swim to the eggs. The fertilized egg germinates into a new, diploid adult. All of this is very much like reproduction in humans.

As an example of a more complex brown life cycle, the kelp fronds that we see on beaches are the diploid sporophyte generation. When fertile, these produce sporangia, in patches called sori, that release meiospores. Kelp sex is genetically determined: half of the meiospores receive a male genetic locus from their diploid parent and germinate into male gametophytes, and half receive a female genetic locus and develop into female gametophytes. The gametophytes are microscopic and filamentous and look very different to the large diploid sporophytes. The female gametophytes produce eggs in oögonia and the males produce sperm in antheridia. When the sperm swims to the egg, retained in the ruptured wall of the oögonium, the fertilized egg

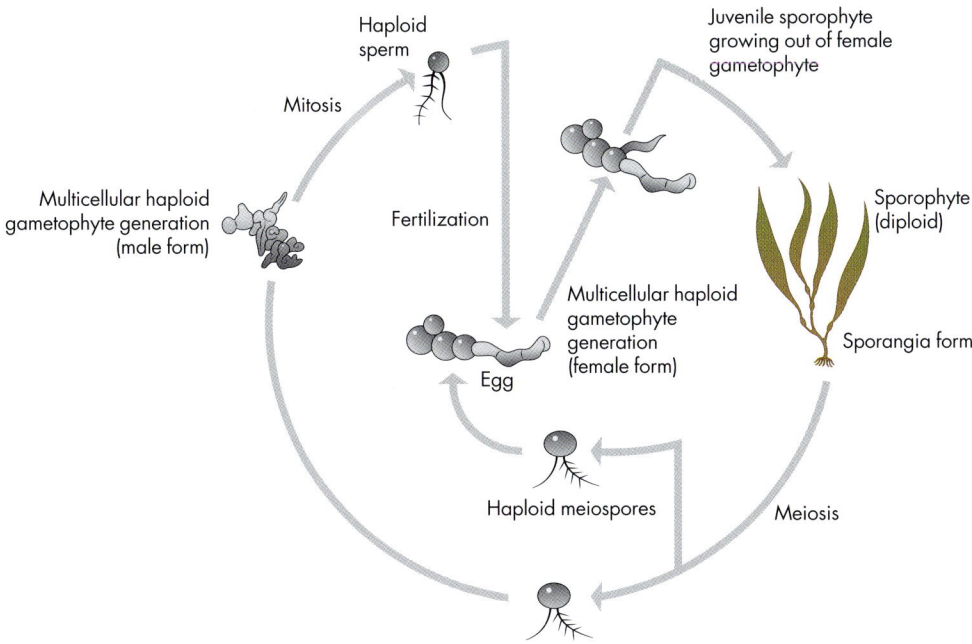

Haploid
sperm

Mitosis

Juvenile sporophyte
growing out of female
gametophyte

Multicellular haploid
gametophyte generation
(male form)

Fertilization

Sporophyte
(diploid)

Multicellular haploid
gametophyte
generation
(female form)

Egg

Sporangia form

Haploid meiospores

Meiosis

germinates to form a new sporophyte, which
grows out of the tiny female parent.

WHY SUCH COMPLEX LIFE CYCLES?

Most seaweed life cycles are so unlike ours that
the temptation is to describe them in detail
without thinking to ask why they should have
evolved. A complication is that some seaweed
species are able to bypass life cycle stages
altogether. The red *Porphyra* of the North Atlantic,
for example, appears to reproduce asexually and
doesn't give rise to a conchocelis phase.

So, we are still debating why seaweeds have
such complex life cycles. For now, the best
explanation is that complex life cycles may provide
more flexibility than simple ones and perhaps help
seaweeds survive in the wide and variable range of
environments that they occupy.

OPPOSITE | The life cycle of the
brown seaweed genus *Fucus* is
essentially the same as ours. The
Fucus species shown is dioecious
and oögamous.

ABOVE | The life cycle of the brown
kelp seaweed is heteromorphic,
dioecious, and oögamous.

SEAWEED PHOTOSYNTHESIS

Photosynthesis is a highly conserved two-stage process, but seaweeds have their own subtle twists on it. The basic reactions of eukaryotic photosynthesis occur in the chloroplast. The inside of the chloroplast is called the stroma and contains a membrane-bound compartment called the thylakoid. The first stage of photosynthesis begins when light hits molecular complexes called photosystems, which are laid out in the thylakoid membrane like an array of solar panels. The photosystems use the energy from sunlight to pull electrons out of the chemical bonds in water, generating oxygen (O_2) and protons (H^+) as they do so. The oxygen diffuses out of the cells, while the electrons and protons flow through a molecular circuit in the thylakoid membrane, eventually passing out of the thylakoid and into the chloroplast stroma. The energy released is used to make two molecules called ATP and NADPH. The second stage of photosynthesis occurs when ATP and NADPH are combined with carbon dioxide in the chloroplast stroma to make sugars: we call this "carbon fixation."

Harvesting light is difficult. The main component of a photosystem is chlorophyll itself, but this can only use red and blue light to split water, which is why plants are green: sunlight contains all the colors of the rainbow but when it hits chlorophyll its red and blue part is absorbed and its green light is scattered back into our eyes. The inability of chlorophyll to absorb green light means that plants miss out on energy from the middle, green wavelengths of sunlight. For this reason, photosystems have developed accessory pigments called carotenoids, which absorb a little more of the green light. Plants arrange their carotenoids around their chlorophyll to act as light funnels: the accessory pigments capture energy that chlorophyll cannot and pass it into the photosystems, allowing photosynthesis to use more of the energy of sunlight.

All of this is complicated in seaweeds because seaweeds live in water. Water does two things: first, it impedes the transmission of light, so seaweeds can struggle to get enough to their photosystems. Second, water is dense, which slows down the

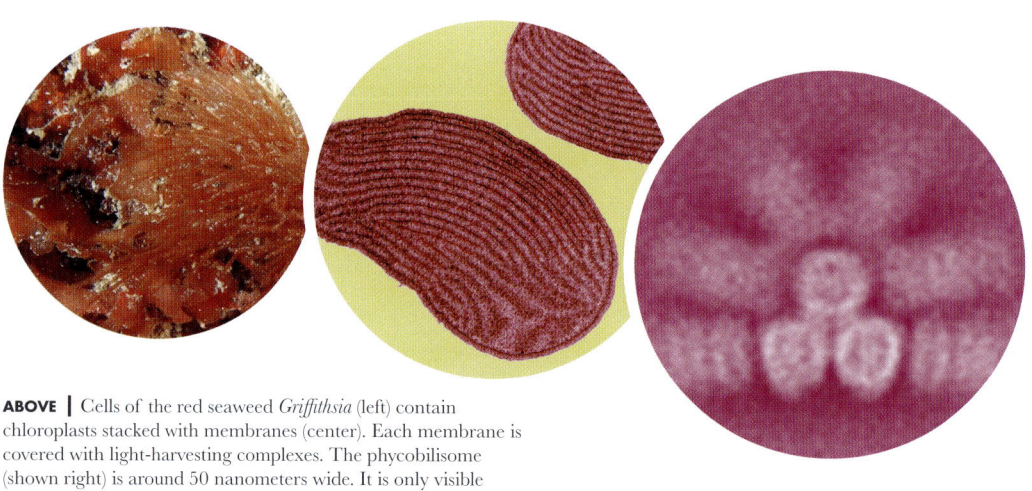

ABOVE | Cells of the red seaweed *Griffithsia* (left) contain chloroplasts stacked with membranes (center). Each membrane is covered with light-harvesting complexes. The phycobilisome (shown right) is around 50 nanometers wide. It is only visible using advanced X-ray microscopy.

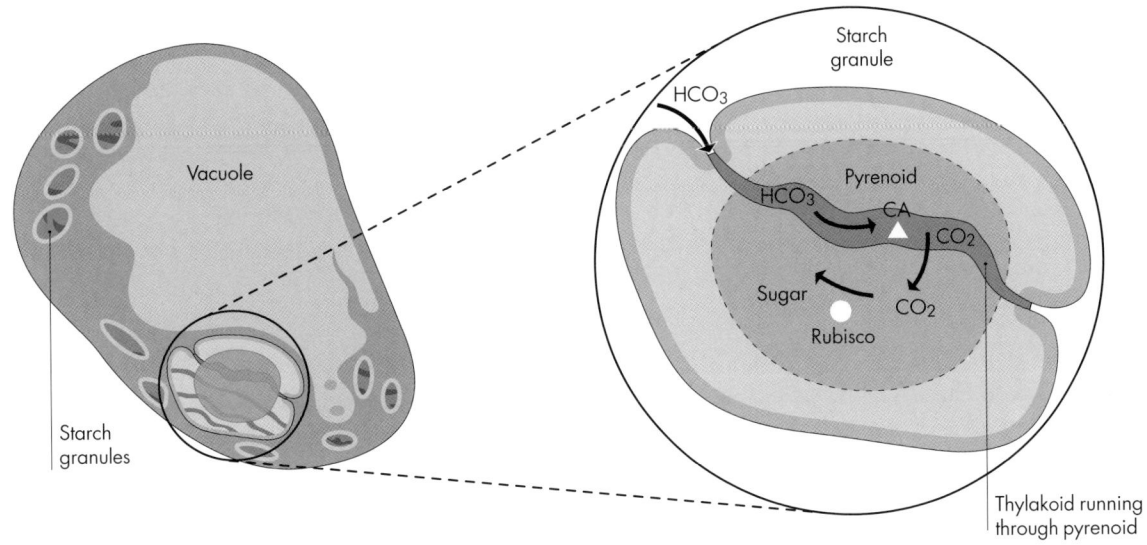

Starch
granule

HCO₃

Vacuole

HCO₃   Pyrenoid
              CA   CO₂

Sugar
          CO₂
      Rubisco

Starch
granules

Thylakoid running
through pyrenoid

diffusion of carbon dioxide to the key chloroplast enzyme, Rubisco, that catalyzes carbon fixation.

Seaweeds solve these problems in two ways. First, they have evolved unique accessory pigments that work well in water. Green seaweeds have one main carotenoid, called α-carotene, but red algae live in deeper, darker waters and have extended their metabolism to make the additional γ- and β-carotenes and the carotene derivatives zeaxanthin and astaxanthin. Many reds have additional pigment-protein funnels called phycobiliproteins, which combine proteins with the pigments phycocyanin and phycoerythrin. These phycobiliproteins are stacked together in intricate structures called phycobilisomes, which may be seen through an electron microscope on the thylakoid membrane of the red seaweeds. The brown seaweeds, having taken their chloroplasts and pigments from the reds, have taken the xanthin pigments one metabolic step further to make a brown pigment called fucoxanthin.

So much for delivering light to photosystems: how do seaweeds deliver carbon dioxide to Rubisco? The answer lies in a cell component called the pyrenoid, which is found in the

ABOVE | The pyrenoid is an algal-specific organelle that helps to supply the chloroplast with enough carbon dioxide to keep photosynthesis running.

chloroplasts of seaweeds but not land plants. In seawater, carbon dioxide reacts with water to form bicarbonate ions (HCO_3^-) according to the following equation: $CO_2 + H_2O \leftrightarrows HH^+ + HCO_3^-$. Around 90 percent of the carbon in water can be locked up in the bicarbonate ion, rather than in carbon dioxide, and this is a problem because Rubisco can only use carbon dioxide as a carbon source. The pyrenoid helps to unlock the carbon in bicarbonate: it is densely packed with Rubisco and sits very close to a part of the thylakoid that contains an enzyme called carbonic anhydrase. The Rubisco in the pyrenoid keeps carbon dioxide levels low by fixing it into sugar. Under those conditions, carbonic anhydrase catalyzes the net conversion of seawater bicarbonate into carbon dioxide, so all the seaweed needs to do is pump seawater bicarbonate into the thylakoid and carbon will be kept flowing from bicarbonate into carbon dioxide and then into sugars.

PRIMARY PRODUCTION

Perhaps the single most important thing to know about seaweeds is that they provide the foundations for life in coastal regions. They do this by creating organic material out of inorganic matter in a process called primary production. The vast majority (more than 99 percent) of the world's primary production occurs through the photosynthesis of plants and algae.

Current estimates put global primary production at around 250 billion tons of new biomass every year. To put that incomprehensibly large number into some sort of perspective, the buildings in a major world city, such as London or Shanghai, are estimated to weigh somewhere between five and ten billion tons, which means that the world's plants and algae build somewhere in the region of 25–50 major cities' worth of biomass every year.

It is hard to estimate how much of this global primary production is down to seaweeds. The best estimates are that land plants and marine algae each contribute about one half of the world's primary production. Most of the algal contribution comes from the unicellular microalgae, the phytoplankton, that live in the oceans' surface waters. Estimates for the contributions of seaweeds are more modest and they contribute perhaps a few percent of the world's annual primary production.

Those few percent are critical, however, because they represent the world's coastlines and there are three particular ecosystems that rely on seaweeds. First, the large offshore forests of kelps and reds that, globally, may represent the largest stands and contributions of the seaweeds. Second, the true intertidal region, which dries and resubmerges with each tide and with which we are most familiar. And, third, the large seaweed rafts that can form floating islands thousands of miles from the shoreline. The best known of these are in the Sargasso Sea, but smaller seaweed rafts are found worldwide.

THE TIMING OF SEAWEED GROWTH

Marine algae grow when they have light and nutrients. The light comes from above, but the nutrients come from below: water is relatively low in nutrients but the seafloor is rich in them. Marine productivity is therefore highest in areas where the seafloor is regularly stirred, bringing nutrients up from the seafloor to be near the sunlight at the sea surface. This happens, most notably, near the poles: surface waters begin to cool in the fall, becoming denser than the warmer, deeper waters below them. The layers of cold surface water move downward, stirring up sediment on the seafloor, especially along the relatively shallow continental shelves. These productive temperate waters support large kelps and, historically, the world's great fisheries: the Bering Sea, the North Sea, the Grand Banks off Newfoundland, and the whaleries of the Southern Ocean. Tropical waters, by comparison, are less stirred and less productive, although often more colorfully diverse.

OPPOSITE | A giant kelp (*Macrocystis pyrifera*) in clear waters near Santa Catalina Island, off the coast of California, USA.

Seaweed primary production is cyclical, varying with the seasons, the tides, and the days. Seasonally, seaweeds begin to grow soon after winter, when the slow mixing of cooling surface waters and warmer, deeper waters brings increasing amounts of nutrients to the upper ocean. Some seaweed species are annual, such as the kelps that grow afresh every year, while others are perennial, like the wracks that lie in the intertidal zone. Some seaweed species are short-lived, with many *Ulva* individuals lasting for a few weeks, while others live for several years.

Seaweed growth also has its rhythms within the seasons. Most obviously, seaweeds in the intertidal zone will slow photosynthesis as the tide goes out and they dry, restarting when the tide returns to cover them with the water that they need to photosynthesize. Seaweeds do not completely stop their metabolism when the tide goes out but produce a range of protective chemicals that are often rich in sulfur and which, when they break down, contribute to the slightly elemental tang of the sea air.

INTERACTIONS WITH OTHER COASTAL LIFE

Seaweeds form the basis of coastal food webs, which means they form habitats and food sources for almost everything else that we find on the seashore. Most familiarly, offshore seaweed stands and forests provide shelter and stability for juvenile fish. Most fish reproduce by releasing their eggs and sperm into the water column, where they mix, fertilize, and settle to grow. It is much harder to ensure that eggs will find sperm if both are released in open ocean, so most fish swim to inshore waters to reproduce and allow their offspring to grow. The mermaid's purses commonly found along Atlantic coastlines,

for example, are the embryo sacs of the ocean- dwelling sharks and rays that were left to develop after their parents had come into shallower waters to breed. Young fish are tempting snacks for older fish, so sheltering them becomes paramount and protection is something that seaweed forests do very well. Estimates are hard to come by but we think that perhaps three-quarters of Earth's fish start their lives in a seaweed bed.

As primary producers, seaweeds are often eaten themselves. Many fish graze on seaweeds, especially the softer reds and greens and on the juvenile stages of the larger browns when they are still little more than a lawn of soft microscopic embryos. A particularly devastating predator is *Strongylocentrotus purpuratus*, the Pacific purple sea urchin, which can clear giant kelp beds off California and which, in its turn, is eaten by sea otters.

Like all living things, seaweeds suffer from disease. They can be infected by other eukaryotes, and the stramenopile oömycetes that cause plant blights have a marine relative called *Eurychasma*, which causes comparable seaweed blights. Seaweeds can suffer from bacterial infections: the crop seaweed *Kappaphycus alvarezii* is prone to outbreaks of "ice-ice," a commercially damaging bacterial disease that may be triggered in warmer weather and can wipe out harvests by rotting the seaweeds in their beds. Finally, seaweeds can suffer from viral infections. The sea is, literally, swimming in viral particles: estimates are that a teaspoon of seawater will contain around one million viral particles, each one perhaps a tiny 100 nanometers across. We know that these viruses infect seaweeds in the lab, but the extent to which seaweed populations are shaped by their diseases remains a much-asked question.

OPPOSITE | A green sea turtle feeding on the brown seaweed *Padina* off the coast of Ecuador.

LEFT | The aptly named seaweed blenny, laying its eggs in seaweed off the coast of Texas, USA.

CENTER | Green seaweeds bleached by the bacterial infection "ice-ice" in Alderney, an English Channel island.

BELOW | The brown *Ectocarpus tomentosus* growing as an epiphyte on *Fucus vesiculosus*, off the coast of Cornwall, UK.

GEOGRAPHICAL DISTRIBUTION

The seaweeds appeared on Earth long before today's continents reached their current locations and the red, green, and brown seaweeds are all found worldwide. However, seaweeds cannot grow everywhere: they need sunlight and nutrients. As a rough rule of thumb, decent amounts of sunlight will penetrate no more than 660 ft (200 m) into clear water and much less than that if the water is cloudy. There are always exceptions to any rule, but these numbers make it unusual to find seaweeds growing any deeper than 330 ft (100 m) in clear tropical waters or deeper than around 80 ft (25 m) in cloudier temperate waters. Broadly, these correspond with the oceanic habitats that we call the intertidal and subtidal zones.

The intertidal zone will be familiar. The continents are defined by their coastlines, which wax and wane with the tides: the area between low and high tide is the intertidal zone. The depth and length of the intertidal zone varies very widely across the different beaches and shorelines of the world. To give two extreme examples, seas that are essentially enclosed basins, such as the Baltic and Black Seas, have tidal ranges that can be measured in inches rather than feet, giving tiny intertidal zones that expose only a few extra inches of the coast. At the opposite extreme, a large spring tide in the Irish Sea can be around 30 ft (10 m) in range. One of the shallow estuaries in northwest England that opens into the Irish Sea is called Morecambe Bay and, when this gently sloping estuary meets a big Irish Sea spring tide, the sea can retreat more than 5 miles (8 km), exposing huge swathes of intertidal mudflats.

The subtidal zone will be less familiar because most of us never get to see it. Our continents sit on platforms called continental shelves, which usually slope gently and gradually out for tens or hundreds of miles offshore before dropping more sharply into the deeper slopes that plummet several miles down to the deep ocean floor. These gently sloping continental shelves are usually around 330–660 ft (100–200 m) deep. The North Sea between Britain and Scandinavia, the Grand Banks off Newfoundland, and the Sunda Shelf between Vietnam and Sumatra are all examples of such continental-shelf seas and the subtidal zone is loosely defined as the shallower regions of these continental shelves.

The intertidal zone uncovers the top few feet of the habitats that seaweeds can live in. The rest of the seaweeds, and the bulk of the world's seaweed biomass, lives underwater. So, anyone standing at the seaside and looking out to sea should remember that the few miles between them and the horizon will hide the subtidal zone and will often be covered in rich and beautiful forests of seaweeds growing on the seafloor.

OPPOSITE | The *marimo* of Japan, the green macroalga *Aegagropila linnaei*. Particular current conditions can roll filamentous green seaweeds up into these distinctive balls.

ARCTIC OCEAN

The Arctic Ocean is the smallest of the world's oceans, covering around 3 percent of Earth's surface. It should be cold and covered in ice. Its waters are less saline than those of other oceans because the Arctic is fed by rivers from a disproportionately large watershed. Average seawater has a salinity of around 35 g salt/kg water (often written as parts per thousand or ‰) but the Arctic has ‰ salinities in the low 30s.

Arctic coastlines are supported by kelp forests (including coldwater species of *Laminaria*, *Saccharina*, and *Saccorhiza*). The understory often contains the brown *Desmarestia aculeata* and species of the red *Phycodrys*. The Arctic intertidal contains northerly populations of small greens (for example, various *Cladophora* species), reds (including *Ceramium*, *Phymatolithon*, *Porphyra*), and brown wracks (such as *Ascophyllum and Fucus*). These are well adapted to their harsh environment: many can shut down photosynthesis for several months during the sunless Arctic winter and live off stored reserves.

Devastatingly, climate change has seen a steady warming of the Arctic since at least the mid-twentieth century, with a corresponding decrease in Arctic sea ice. Ice reflects light, so as ice retreats, more light is absorbed and more sea is warmed. This feedback has made the Arctic the fastest-warming part of the planet. Established seaweed and animal species are likely to die and new species are expected to to move in as the Arctic warms and its channels open.

SOUTHERN OCEAN

The Southern Ocean is generally defined as the waters below 60°S and covers around 4–5 percent of Earth's surface. Its waters are cold and average around 34‰ salinity, saltier than the Arctic but not as salty as the three oceans that sit above it. The Southern Ocean is defined by a current—the Antarctic Circumpolar Current—rather than by landmasses and this current helps to supply upwelled nutrients from the seafloor, giving the Southern Ocean some of the world's most productive waters.

ARCTIC OCEAN

SOUTHERN OCEAN

 Arctic Ocean

■ Southern Ocean
⇨ Antarctic Circumpolar Current

Much of this productivity comes in the form of oceanic phytoplankton, rather than coastal seaweeds, but research stations are sited on the relatively hospitable Antarctic Peninsula to the south of Argentina and we know that its subtidal zone is dominated by many large species of *Desmarestia*, including the aptly named *D. antarctica*. The large *Himantothallus grandifolius* is also prominent. This looks very like the kelps of higher latitudes, but is more closely related to the nearby *Desmarestia* species. Some cold-adapted greens (for example, *Monostroma* and *Ulva*) and reds (such as *Sarcopeltis*, and *Phycodrys antarctica* and *austrogeorgica*) have also been reported, sometimes extending along the Antarctic coastline to the western side of the Ross Sea, roughly south of New Zealand.

ATLANTIC OCEAN

The Atlantic covers around 17 percent of Earth's surface and has the highest mean salinity of any ocean, ranging from 33–37‰. Its coastlines are set beside continental-shelf seas, most of which extend 50–100 miles (80–160 km) offshore before dropping into deep basins that sink to the sunless ocean floor. Atlantic coastlines often show seaweed zonation, with small green seaweeds at the top, belts of brown wracks and red algae in the intertidal zone, and brown kelp forests interspersed with red and green algae in the subtidal zone.

Seaweed species differ between the North and South Atlantic. The oceans are connected by currents that flow into each other in cycles called gyres. These are driven by Earth's rotation, flowing clockwise in the northern hemisphere and counterclockwise in the southern hemisphere. Their opposing flows keep southern and northern hemisphere waters and species reasonably separate.

So, much of the North Atlantic intertidal is dominated by the branched Fucoid wracks, with serrated wrack (*Fucus serratus*) and bladderwrack (*Fucus vesiculosus*) two of the easier species to spot. Larger laminarian kelps, the oarweeds (*Laminaria*) and sugar kelps (*Saccharina*), can extend hundreds of feet offshore in the northern subtidal zone, with an understory of reds. The kelps also dominate in the South Atlantic subtidal zone, but the genera are now *Lessonia* and *Macrocystis* off the coasts of South America, and *Ecklonia* and the species *Laminaria pallida* off the coasts of South Africa. South Africa, in particular, has diverse red understories growing in its *Ecklonia* forests: *Gelidium*, *Gracilaria*, and *Porphyra* species are common.

OPPOSITE, LEFT | Japanese wireweed, the brown seaweed *Sargassum muticum*, in the English Channel off Brittany, France.

OPPOSITE, RIGHT | At the other end of the Atlantic, a kelp canopy off the coast of South Africa.

ATLANTIC OCEAN

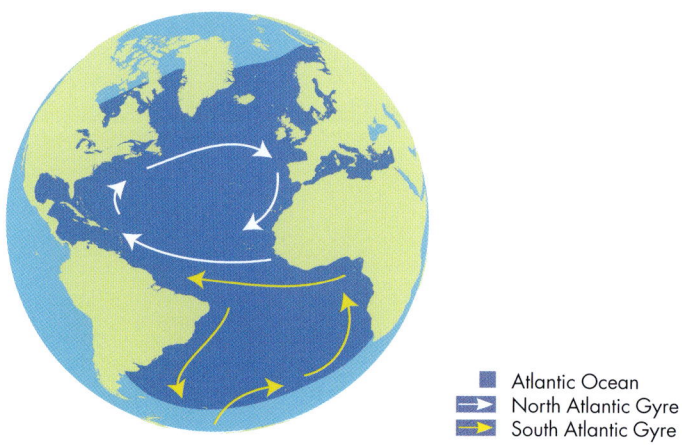

■ Atlantic Ocean
➡ North Atlantic Gyre
➡ South Atlantic Gyre

MARGINAL SEAS OF THE ATLANTIC

The Atlantic's marginal seas have drastically different characteristics: the Baltic is fed by river basins from the mountains of Finland and Sweden and is cold and shallow, with a salinity that rarely rises above 10‰. It is an extraordinary natural laboratory whose seaweeds have adapted to growing almost in freshwater.

The warmer, deeper, and saltier waters of the Mediterranean are not as nutrient rich as those of the Atlantic and do not support kelps. Instead, Mediterranean shorelines see more delicate browns, such as the bushy *Cystoseira* and the pottery-like *Padina*. Reds are common and distinctive, with well-established maerl beds. Greens are also well represented, although green *Caulerpa* species and the red *Asparagopsis* have entered through the Suez Canal from the Indian Ocean and have become invasive.

On the western side of the Atlantic, the tropical waters of the Caribbean support smaller browns, such as *Dictyota*, but are noted mainly for the beauty and diversity of their red and green seaweeds. The many delicate species of *Caulerpa* and siphonous greens such as *Cladophora* are a Caribbean highlight.

PACIFIC OCEAN

The Pacific is enormous, covering around one third of the world's surface. It is generally fresher and deeper and colder than the Atlantic. As in the Atlantic, the North and South Pacific gyres keep northern and southern hemisphere seaweeds apart. The North American Pacific coast harbors offshore forests of the giant kelp *Macrocystis prolifera*, the elk kelp *Pelagophycus*, and the bull kelp *Nereocystis luetkeana*. Smaller kelps such as *Egregia* are common closer inshore, as are fucoids. The Asian side of the North Pacific is also dominated by offshore kelps, especially from the genera *Ecklonia*, *Eisenia*, and *Undaria*. Reds such as *Gelidium* form understories on both sides.

In the south, the coastlines of Chile and Peru also support huge offshore kelps, both *Macrocystis* and the southern hemisphere *Durvillaea*. These coastlines represent a notable exception to the continental shelves seen elsewhere in the world: the sea drops precipitously into the Peru-Chile Trench, reaching depths of around 4 miles (6.5 km) within a day's sail of shore. The Peru Current flows northward through this trench, bringing with it cold, nutrient-rich water from the Southern Ocean. These nutrient-rich waters support not only the kelps, but also diverse red understories, including the *Sarcopeltis* genus, which has some of the largest thalli of any red seaweed. The South Pacific Gyre means that the kelp genera *Durvillaea* and *Macrocystis* are also found on the shorelines of New Zealand almost 6,000 miles (10,000 km) away, which they share with the brown *Ecklonia* kelps and the distinctive brown *Hormosira*, along with many species of reds (*Osmundaria*, *Porphyra columbina*, and *Gigartina*).

MARGINAL SEAS OF THE PACIFIC OCEAN

The Bering Sea, the Sea of Okhotsk, and the Sea of Japan sit in the colder, fresher North Pacific and support its characteristic kelp forests. The Pacific's western border, however, is bounded by a chain of tropical seas that run from the Coral Sea off northeast Australia, through the Arafura and Banda Seas around Indonesia, up through the Celebes and Sulu Seas to the west of the Philippines, and on through the South China Sea and East China Sea, and the Yellow Sea back to the Sea of Japan. These shallower tropical seas are home to a huge variety of red and green seaweeds.

PACIFIC OCEAN

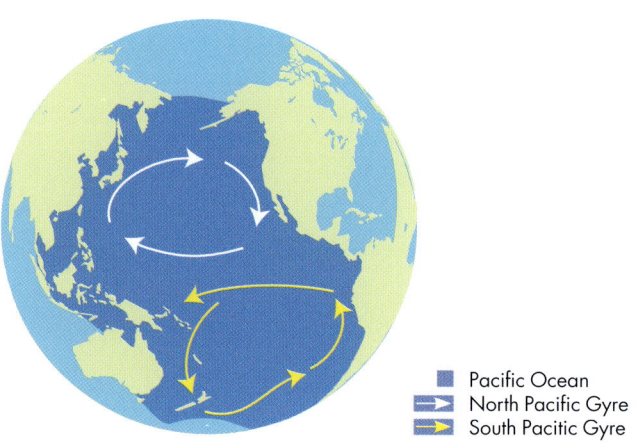

Pacific Ocean
North Pacific Gyre
South Pacific Gyre

The reds *Eucheuma*, *Gracilaria*, and *Kappaphycus* are common and farmed, especially in the Philippines and Indonesia, and *Neopyropia yezoensis* farms are found north of Shanghai. The red genera *Jania*, *Asparagopsis*, *Botryocladia*, and *Ceratodictyon* are found near coral reefs in the region, and the green *Valonia*, *Caulerpa*, *Halimeda*, and *Cladophora* genera are also common. Many smaller browns are also seen, including the tufty *Cystoseira* species, the spikier *Turbinaria*, and the ever-present *Sargassum*.

ABOVE | The brown seaweed *Durvillaea* on rocky shorelines in Doubtful Sound in the southwest of New Zealand's South Island (Southland/Murihiku region).

LEFT | A harvest of the red seaweed *Kappaphycus* on Kangge, one of southeast Indonesia's lesser Sunda Islands.

INDIAN OCEAN

The Indian Ocean covers about 14 percent of the Earth's surface. On average, it is as saline as the Pacific, 32–37‰, but there are sharp divsion between its salty northwest and its comparatively fresh northeast, where the Bay of Bengal is diluted by the vast watersheds of the Ganges-Brahmaputra-Meghna and Myanmar's Irrawaddy. Temperatures are often in the 80s (°F) in the northern Indian Ocean, dropping to around 32°F in its lower latitudes where it meets the Southern Ocean. Like the Arctic, the Indian Ocean is visibly suffering from the effects of climate change: its sea levels are rising, its corals are dying, and its surface waters are warming.

Currently, species from the brown *Turbinaria* and *Sargassum* genera are found throughout, but the kelps of the South Atlantic do not grow well in the warmer waters of the Indian Ocean and do not continue east around Cape Agulhas.

Instead, large populations of the reds *Hypnea* and *Gelidium* are found on the east coast of South Africa and *Eucheuma* is farmed along the coastlines of Kenya and Tanzania. Red *Gracilaria* and *Acanthophora* species are also common along south

India and Sri Lanka, running down Indonesia to western Australia. Of particular interest are several red genera, such as *Bostrychia*, which grow as epiphytes on the roots of the mangroves in Tanzania and Mozambique. The waters of the Indian Ocean are also rich in green seaweeds, especially in the equatorial waters of the northern Indian Ocean: *Ulva* and *Caulerpa* species are found along the shores of Gujarat and the distinctive *Valonia* is found along the coasts of South Africa, Kenya, and Tanzania.

MARGINAL SEAS OF THE INDIAN OCEAN

The Indian Ocean stretches from the Great Australian Bight at its southeast, running around the western side of Australia and up through Indonesia to include the Timor, Flores, and Java seas. These are part of the same chain of shallow tropical seas that form the boundary of the western Pacific and have similarly rich red and green seaweed flora: the green *Caulerpa*, *Cladophora*, and *Codium* genera are common, as are a host of reds such as *Plocamium*, *Grateloupia*, and *Griffithsia*, and many more. The Indian Ocean's fresher northeast takes in the Andaman Sea and the

INDIAN OCEAN

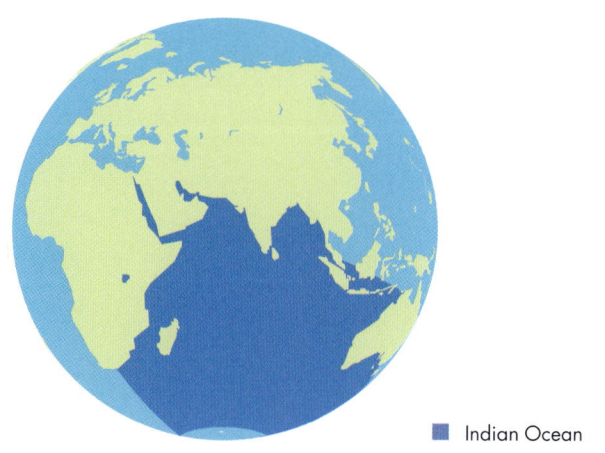

■ Indian Ocean

Bay of Bengal, and its saltier northwest sees the Arabian Sea open into the drier, more saline Persian Gulf and the Red Sea. The brown *Dictyopteris* and *Padina* are common in these warmer northern waters, as are the red *Asparagopsis* and the green *Caulerpa* species that have invaded the Mediterranean.

RIGHT | The calcifying red seaweed *Jania rubens*. This species is relatively common to the north and west of the Indian Ocean.

BELOW | The red seaweed *Gracilaria aculeata*, found in the Indian Ocean off the coast of KwaZulu-Natal, South Africa.

Food, Fodder, and Fertilizer

FOOD, FODDER, AND FERTILIZER

Coastal communities have been eating seaweeds for thousands of years. The more delicate species may be eaten raw. Dulse, for example, the leafy red seaweed, *Palmaria palmata*, tastes a little like salami—it is chewy, salty, and slightly peppery.

Seaweeds like dulse have been eaten in this way for centuries. A common division of labor in preindustrial fishing villages saw the men head out to sea to fish, while the women waded out to collect fresh "sea vegetables" to go with the fish that the men would bring home. The nineteenth-century British periodical *Household Words* describes "dulse-wives" selling their wares in the port of Aberdeen (in the following, a "creel" is a fisherman's basket and "sonsie" is a Scots word that means jolly): "They sat in a row on little wooden stools, with their wicker creels placed before them on the granite paving stones. Dressed in clean white mutches, or caps, with silk-handkerchiefs spread over their breasts, and blue stuff wrappers and petticoats, the ruddy and sonsie dulse-women looked the types of health and strength."

Despite this long history, seaweed is not everybody's idea of a good meal. Writing in 1902, the American phycologist William Setchell wrote of Hawaii: "The foreigner, as a rule, hesitates to eat raw seaweed, but when he tries the unaccustomed food he finds a variety of flavors and a relish in some of the species which amply repays him for his courage in making the attempt." It took one of Setchell's collaborators, the pioneering Chinese phycologist Cheng Kui

Tseng, to change this, building up traditional mariculture into his "farming and ranching of the sea" that has now become a multimillion-dollar industry feeding millions of people worldwide.

Seaweeds compare well nutritionally to most of the vegetables that we eat today, are often higher in protein than land vegetables, and are comparably low in fat. They are also relatively rich in certain vitamins, particularly folate (B9) and riboflavin (B2), and in nutrients such as iodine. Seaweed carbohydrate is slightly different to that of land plants, so seaweed will probably be a little less digestible if it is not a regular part of the diet. However, in humans the dietary benefits of plant fiber depend critically on the exact composition of both carbohydrate and our gut bacteria. Human populations that regularly feed on seaweed have intestinal microbiota that are more efficient at breaking down seaweeds. People in Japan and coastal Spain, for example, have gut bacteria that allow them to digest seaweed carbohydrates more efficiently than humans who live farther inland.

COOKING WITH SEAWEED

Most seaweed is not eaten raw but is prepared and cooked. Prior to cooking it is often dried and may also be fermented. Many communities have their own traditions for turning crops of tough seaweeds into edible sea vegetables, although the underlying techniques are much the same worldwide. The seaweeds are often drip-dried to remove as much seawater as possible, before being washed or boiled. The cleaned and chopped seaweeds can then be pressed into flat sheets. This is the technique used in Japan to make the *nori* that is wrapped around sushi. In communities around the Yellow Sea, sheets of washed and

OPPOSITE | A *Kappaphycus* farm in Bali, one of southeast Indonesia's lesser Sunda Islands. The seaweeds are grown on ropes, which are visible in the foreground.

dried seaweeds are flattened and cut into wafers and sold in packs as a snack food. An alternative approach is to boil or blanch raw seaweeds to create a preserve. This method is commonly used for smaller species and is the basis of both the Hawaiian *limu* cuisine (*limu* is the Polynesian word for edible marine plants) and of several Japanese dishes, most notably wakame and kombu, which are based on preserved or pickled kelps.

FODDER

Humans are not the only land animals who eat seaweeds. Many coastal creatures will, although usually only as a last resort. Arctic foxes have been seen to eat them in the absence of anything better to feed on and Australian ruminants will graze on seashore kelps. The most studied examples are the famous North Ronaldsay sheep. North Ronaldsay is the northernmost of the Scottish Orkney islands and looks directly at the North Pole across 2,000 miles (3,200 km) of sea. The grazing on land is poor, so the sheep have acquired a distinctive set of gut bacteria that can break down seaweed biomass. This allows them to graze almost entirely on seaweed for large parts of the year.

Seaweed has long been used to supplement animal fodder, especially in coastal areas around the rocky shorelines of northwestern Europe. For thousands of years, farmers in Ireland, Scotland, and Scandinavia have fed silage from kelps and wracks to their cattle, pigs, and sheep and a number of companies now sell milled seaweed as an animal fodder supplement. Many studies on the use of seaweed as animal fodder were carried out during the wars of the twentieth century, when various European countries were more reliant on growing their own food. The studies show that seaweed is a perfectly good fodder, limited mainly by being in competition with cheap land crops.

Those economic considerations have been cast in a new light by climate change. Climate change is driven by human-produced "greenhouse gases," which are so called because their effect is to make Earth's atmosphere behave like a greenhouse. Carbon dioxide (CO_2) is the best known of these greenhouse gases because it is released in the largest quantities, but methane (CH_4) is as important: we release more carbon dioxide, but methane is a more potent greenhouse gas and our best estimates are that both gases are equally dangerous and equally in need of reduction. Methane is produced by various bacteria as a side effect of fermentation. So, when we farm ruminants, we are also farming their gut bacteria and those gut bacteria produce methane in massive quantities. Estimates suggest that agricultural methane may contribute around one fifth of our greenhouse gas emissions.

Seaweed carbohydrates are not digested in quite the same way as land plant carbohydrates. Several recent studies, most notably using the red seaweed *Asparagopsis*, have suggested that for this reason cows fed on seaweed fodder produce a lot less methane than cows fed on their normal meal. This research is still at its very early stages but holds real potential for considering how we can use our natural resources more sustainably.

FERTILIZER

Rotted down seaweeds make excellent compost. They have, therefore, a long history of use as fertilizer in rocky coastal areas, where seaweeds are abundant but soils are often thin. It has been argued that long, stone troughs built almost a thousand years ago by Viking settlers on shorelines around the northeast Atlantic were used as kelp compost silos to provide the fertilizer that would allow settlers to grow familiar land crops. The heavier brown seaweeds are rich in nutrients that are released as the thalli rot and some species, most notably *Ascophyllum*, release more complex compounds that can stimulate plant growth in much the same way as the plants' own hormones. The famous artichokes of Brittany used to be planted in soils fertilized by kelps collected from its coastlines. The use of seaweed fertilizer was so important in these areas that seaweeds were farmed if they could not be harvested. Well into the twentieth century, stone piles were laid out on sandy beaches along the west coast of Ireland to give wracks and kelps something to attach to and grow on. The seaweed that grew naturally there was then gathered and used as fertilizer.

BELOW LEFT | Seaweed-grazing sheep on North Ronaldsay in the Orkneys, Scotland.

BELOW RIGHT | Sorting seaweed in Maine, USA.

EXTRACTIVE INDUSTRIES

Some seaweed components are worth more than others. Multiple industries have been created to extract and purify these higher-value components from the biomass of various species. There are two classes of high-value products that may be extracted: inorganic salts and elements (such as potash, saltpeter, and iodine) and biopolymers (for example agar, carrageenan, and alginate).

POTASH

Living things contain valuable elements, which means that dead things can provide those elements, too. Around 1.5 percent of your dry mass is made up of calcium, for example, and 0.5 percent is potassium. These elements are not present in their pure elemental forms, but as salts or in molecules, and potash is one of those salts. Its name means, literally, "pot ash" because it was made by soaking wood ash in water. The soluble potash dissolves into the water and is recovered by allowing the water to evaporate, leaving potash salt as a residue. The original word was actually Dutch: *potaschen*, but the English is cognate.

Potash is valuable because, chemically, it is a collection of potassium salts. The main one is potassium carbonate (KCO_3), historically known as soda. Soda was, and remains, an important industrial chemical, especially in the glassmaking industry where it is used to lower the melting temperature of silica, allowing glass to be blown at temperatures that may realistically be reached by a relatively small furnace.

Potash has also been used as a chemical fertilizer for several hundred years. Growing plants extract potassium from the soil, which means agricultural soils are relatively low in that element. Crop yields may be boosted by seeding soils with more potassium and one of the cheapest ways to get that potassium is in the form of potash.

Today, most of the potash we use is mined but before the industrial era, the main method of making potash was from wood ash. However, wood is slow growing and has other uses. So, when wood was scarce, people often turned to seaweeds. The kelps, especially, may be burned to give potash-rich ashes. Historically, therefore, many preindustrial coastal communities have collected and burned seaweeds to make potash for their industries and agricultures.

SALTPETER

Besides potassium carbonate, potash also contains significant amounts of potassium chloride (KCl), potassium sulfate (KSO_4), and potassium nitrate (KNO_3). The last of these, potassium nitrate, can be as valuable as soda, because, under the name of saltpeter, it is the main component of gunpowder. The NO_3 part of saltpeter is effectively an oxygen store that allows the gunpowder to burn explosively. Saltpeter is usually mined in its mineral form, in which it is known as niter. However, niter mines are uncommon and in times of war countries may be cut off from their saltpeter supplies. This has happened on two

notable occasions over the past couple of centuries. The first was during the Napoleonic Wars of the nineteenth century and the second was during World War I, both of which saw combatants making gunpowder from seaweed.

IODINE

Revolutionary and Napoleonic France used a lot of gunpowder. Its kings, revolutionaries, and emperors drove French chemists to find new ways of making the saltpeter they needed. In doing so, they found a new element and a new industry.

The element was iodine and it was discovered when the French chemist Bernard Courtois saw a beautiful violet vapor appear after he added sulfuric acid to kelp ash. Seaweeds are rich in iodine, which they accumulate from seawater and use as part of their defense mechanisms. It was soon realized that iodine has economic and medical applications: iodine is a strong dye, tinctures of iodine will clean wounds, and iodine

ABOVE LEFT | Potash fertilizer is a rich source of potassium, which is especially beneficial for flowering and fruiting plants.

ABOVE | A long-disused seaweed pit. Seaweed was burned and the iodine-rich liquid bubbled out and was collected after it had run down into the pits.

is also an essential element that our thyroid glands need to make a hormone called thyroxine.

At the start of the nineteenth century, therefore, the economic importance of iodine created a new seaweed-based industry, which supported many poorer coastal communities. British and French crofters would harvest thousands of tons of kelps and process them, as a precursor to today's biorefineries. Kelps were burned in kilns close to shore, the iodine-rich oil that dripped out was collected, and the ashes were used to make potash. However, although kelps gave birth to iodine, the seaweed iodine industry no longer exists. Its heyday lasted perhaps until the mid-nineteenth century, by which time iodine was being extracted from its mined mineral salts.

AGAR-AGAR AND AGAROSE

Seaweed cells are wrapped in elegantly woven cell walls, which are made from long, unbranched, and insoluble cellulose fibers that are "glued" together by shorter, more branched, soluble polymers. In plants these shorter, glue-like polymers are called the hemicelluloses and the

BELOW | In many countries, seaweed farms are young industries that give new economic power to women. These Indonesian women are tying nursery-grown *Kappaphycus* cuttings onto the ropes that will then be staked out to grow in the sea.

RIGHT | Petri dishes with a layer of agar in the bottom. These "agar plates" are used to grow microbial cultures.

pectins. Land plant pectins will be familiar to cooks because they are used to make jam: dried pectin absorbs water and glues together the fruit in the jam in much the same way that it binds together the cellulose fibers in the plant cell wall.

The seaweeds use similar biopolymer glues, which differ between the reds, greens, and browns. Called hydrocolloids, they display similar gel-forming properties to pectin and the animal

product gelatin. Seaweed hydrocolloids are easy to prepare, have historically been used for a range of domestic and industrial applications in Asia and Europe, and are used worldwide in industries.

The dominant biopolymer in many species of red seaweeds, most notably *Gracilaria* and *Gelidium*, is called agar, so those species are correspondingly known as agarophytes. Agar's gel-forming properties have been known since at least the sixteenth century: it has a relatively high melting point, so can be used to form firm jellies at room temperature and is often used across Japan and South Asia to make sweetmeats or desserts, especially as a vegetarian alternative to gelatin.

Chemically, agar is a heterogeneous polymer that may be separated into two main components called agaropectin and the more homogeneous agarose. The homogeneity of agarose means that its gel-forming properties may be tightly controlled, making it useful for a range of scientific applications. Agarose gels, for example, are used in a molecular biology technique called gel electrophoresis, in which DNA fragments of different sizes are separated by using an electrical current to push them through the gel. Smaller DNA fragments will move more quickly through the gel while larger fragments move more slowly, which helps researchers to purify specific genes.

CARRAGEENAN

Agar is not the only kind of biopolymer made by red seaweeds. A different, though related, biopolymer is called carrageenan and dominates

ABOVE | Drying, cleaning, and packing a *Kappaphycus* harvest on Kangge, one of southeast Indonesia's lesser Sunda Islands.

in the red species known as the carrageenophytes: *Gigartina*, *Chondrus*, *Kappaphycus*, and *Eucheuma*. As with agar, carrageenan is easy to extract from seaweeds: both agar and carrageenan are water soluble, so can be extracted by boiling the seaweed biomass and straining off the now carrageenan-containing liquid. Carrageenans have been used as commercial gelling agents since at least the fifteenth century and may be efficiently purified at industrial scales.

Carrageenan is structurally related to agar and is also heterogeneous. It is constructed from three basic building blocks, the monomers α-galactose, β-galactose, and 3,6-anhydro-α-galactose. As their names suggest, these are all variations on the sugar galactose, which has a very similar structure to glucose. These three galactose-based monomers may be combined in different orders and their side chains may be modified to different extents by the addition of sulfate groups to give a range of carrageenans that are distinguished using different Greek letters: α-carrageenan, β-carrageenan, γ-carrageenan, and so on. The variations in monomers and the extent of sulfation affect the gelling properties of the carrageenan, which means that different kinds of carrageenan have different commercial uses.

κ-carrageenan, for example, is built from repeating blocks of β-galactose-4-sulfate and 3,6-anhydro-α-galactose and contains one sulfate group for every disaccharide unit. This structure allows κ-carrageenan to form strong, hard gels, so it is commonly used in toothpaste. λ-carrageenan,

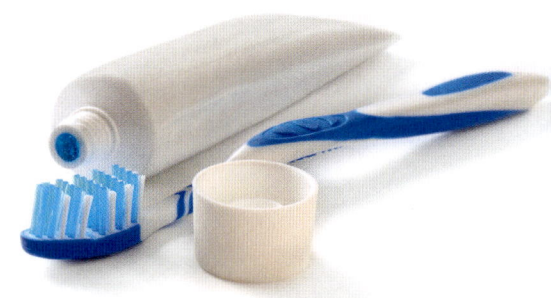

ABOVE | Different kinds of carrageenan can give different kinds of gels: the wobbly gel of the jelly sweets (left) will use a different carrageenan to the softer gel of the toothpaste.

on the other hand, is built from a different pair of repeating blocks: β-galactose-2-sulfate and α-galactose-2,6-disulfate, which means that λ-carrageenan contains three times as much sulfate per disaccharide as κ-carrageenan. The difference in sugars and sulfates means that λ-carrageenan does not form true gels at all but can act as a thickener. λ-carrageenan is therefore often added to plant milks, such as soy milk, to thicken them and make them feel more like animal milks.

ALGINATE

The final major seaweed biopolymer is derived from brown seaweeds. The cell walls of many laminarian seaweeds are rich in a polymer called alginate, which performs much the same glue-like role in the browns as agar and carrageenan do in the reds and as pectin does in land plants. Alginate is made from repeating disaccharide blocks of β-mannuronic acid and α-guluronic acid and is extracted in a similar fashion: the seaweeds are dissolved in hot solvents and the resulting soluble polymers are strained off.

As with the agars and carrageenans, alginate has a range of commercial applications. Alginate polymers have zigzagging molecular backbones, so when two alginate molecules lie side by side,

diamond-shaped pockets are formed as the backbones repeatedly zig toward and zag away from each other. These pockets are about the same size as many metal ions and when metals slot into these alginate pockets, they can form molecular bonds that pull the alginate polymers together. This is called the "eggbox" model because the metals in their alginate pockets look like eggs in an eggbox. Critically, the eggbox model means that both hard gels (for example, dental molds) and soft gels (for example, ice-cream thickeners) can be made from alginate simply by combining the alginate with varying amounts of different metal salts. In practice, alginate and carrageenan are used almost interchangeably in European, American, and Asian cuisines. Soft gels and hard gels can be made from both by changing the type of polymer (carrageenan) or the metal ion concentration (alginate).

ABOVE | Harvesting kelp in Japan.

RIGHT | An alginate-based face mask. Alginate itself is a white powder and provides the gel that holds the rest of the face mask's ingredients together.

BIOFUEL

The simplest way to use complex biological polymers is to burn them for heat. This, after all, is what humans have been doing with wood for thousands of years. Raw seaweed is a less satisfactory fuel than wood but, historically, has sometimes been used in that way by coastal communities. Increasing efforts are now looking at whether seaweed biomass can be used for higher-quality biofuels.

Broadly speaking, there are four kinds of biofuel: the fats in biomass can be burned as biodiesel; the sugars in biomass may be fermented to give bioethanol; the biomass itself may be biologically or thermochemically digested to give biogas (usually methane and hydrogen); or the biomass may be dried to give biochar. All of these approaches are perfectly possible with seaweeds.

Unfortunately, very few of these approaches are currently economically viable. The world does not harvest that much seaweed and seaweeds are usually worth more if they are used for something other than fuel. For example, one of the world's foremost seaweed industries is carrageenan, which harvests perhaps 1 million tons (dry weight) of red seaweeds every year. In the same 12 months, the world uses around five billion tons of oil: we would need to scale up our seaweed yields by 5,000-fold to get anywhere near and the effect on marine ecosystems is unimaginable. Similarly, fermenting biomass to bioalcohol produces a fuel that currently sells for around $4 per gallon. Diluted to make a drink, it will sell for ten times as much. A farmer will make more money from their land (or their sea) if they use it to grow food rather than fuel, which is one of the reasons why biofuel crops are only usually grown when subsidies are on offer.

RIGHT | Sodium alginate beads. Alginate's gel-forming properties make it a popular way to create soft capsules like these, to which green coloring has been added.

OPPOSITE | Raceway ponds, used in laboratories around the world to grow floating micro- and macroalgal species in bulk.

There is some hope in the seaweed biofuel story. First, there is a place for seaweed biofuel in a balanced transition from a petrochemical to a sustainable economy, and this is especially true for aviation fuel. Airplanes need to carry their fuel with them in a way that houses do not, which means they need an extremely energy-dense fuel. Biofuel is several times more energy dense than current electric batteries and is already being trialed as a drop-in fuel for airline fleets.

Second, most of the world's fuel is consumed by higher-income countries. There is significant potential for lower-grade biogas generation to provide power in lower-income countries. Third, and finally, many parts of the world are seeing harmful seaweed blooms: they are, in effect, being given free biomass that they do not want. Free biomass significantly shifts the point at which biofuel generation becomes economically viable and efforts are underway worldwide to turn harmful blooms, such as the Caribbean's annual *Sargassum* tides, into useful biofuel.

SEAWEEDS IN THE ANTHROPOCENE

Geological time is divided into eons (roughly, billions of years), which are subdivided into eras (hundreds of millions of years), which are in turn subdivided into periods (tens of millions of years), which are finally subdivided into epochs (millions of years). *Tyrannosaurus rex*, for example, lived in the Late Cretaceous epoch (101–66 MYA) of the Cretaceous period (145–66 MYA) of the Mesozoic era (252–66 MYA) of the Phanerozoic eon (~540 MYA–now). Eons are very long: we are still living in the Phanerozoic eon, the same eon in which *T. rex* lived, but we have moved from the Mesozoic era into the Cenozoic (66 MYA–now). We have made our way into the Cenozoic's Quaternary period (~2.6 MYA–now) and into an epoch that many now call the Anthropocene: the human-made epoch.

The Anthropocene is so named because, over the past few hundred years, human activity has changed the climate of the world around us. By the early 2020s, atmospheric carbon dioxide levels were around 415 parts per million (ppm), a 50 percent increase over the preindustrial levels that sat around 280 ppm. Carbon dioxide is a greenhouse gas that causes atmospheric warming but it has another, less widely known, effect on the oceans. Carbon dioxide reacts with water to form the bicarbonate ion and protons: $CO_2 + H_2O \rightleftharpoons H^+ + HCO_3^-$. As carbon dioxide rises, the number of protons in the sea also rises and ocean pH drops. Over the past 200 years, oceanic pH is thought to have dropped by around 0.1 pH unit, from around 8.2 to around 8.1. This may not sound a great deal, but the pH scale used to describe acidity is a logarithmic one rather than a linear one, which means that a 0.1 decrease in pH represents around a 20 percent increase in the

ABOVE | Coral bleaching on the Great Barrier Reef, off Queensland, Australia. Corals are a symbiosis of spongelike animals and single-celled algae called dinoflagellates. Warming and acidifying waters can break this symbiosis, resulting in bleaching.

LEFT | An aerial view of offshore kelp forests at Santa Catalina Island, California, USA. These forests are dying back as climate change alters our oceans.

number of acidic protons. So, our oceans are not only warming, they are also acidifying. These rising temperatures and falling pH levels are changing seaweed populations. For example, the kelp forests that drive coastal ecosystems are moving northward and the Pacific kelp forests are dying back at a frightening rate.

A second anthropogenic change involves the destruction or creation of new habitats. The dredge fishing that is often used to gather shellfish can strip seaweeds clean away from the seabed, but this is perhaps less damaging than the inexorable growth of human coastal settlements, whose concrete wharves and breakwaters remove the rocks and sands on which seaweeds used to grow. Habitat creation is also a problem: for example, fertilizer runoff can create offshore

bioreactors for genera such as the green *Ulva* and the brown *Sargassum*. Blooms of these genera can deposit well over a million tons of biomass along a few miles of shoreline, costing several million dollars to remove.

INVASIONS AND HARMFUL BLOOMS

There is a third way in which human activity is changing our coastlines. Until relatively recently, oceanic gyres have kept species separate, but

ABOVE | The invasive green seaweed *Caulerpa taxifolia*.

ABOVE LEFT | Biofouling on the bottom of a boat's hull can raise fuel use by almost 50 percent. The green slime is a mix of micro- and macroalgal species.

humans are now creating new ocean currents that bring species together. In the past two centuries, humans have physically opened two major new waterways. The isthmus of Panama closed around four million years ago, but in 1914 the Panama Canal reconnected the Caribbean to the Eastern Pacific. Similarly, the Mediterranean Sea was created perhaps three million years ago as Africa swung into Asia to create the isthmus of Suez. In 1869 the Suez Canal reconnected the Mediterranean with the Indian Ocean, allowing the entry of the now-invasive *Caulerpa* that annually clogs the waters of the Mediterranean.

Humans have also created shipping lanes that carry species across thousands of miles of ocean, either as fouling attached to the hulls of ships or, less noticeably, as microscopic spores in the ballast water. Stand at any commercial dock and watch the water being pumped out of a container ship and you will be watching seaweed spores finishing a journey that may have moved them from one side of the world to the other. The old seaborne spice route from India to Europe, for example, cuts across three separate oceanic gyres and ships taking the same route are thought to have brought the now-invasive wireweed, *Sargassum japonicum*, to the European waters it is increasingly choking.

SPECIATION

The world is changing and the seas are changing with it. Evolution works by selecting species that survive in their environments, so new species arise as environments change. Speciation, the creation of new species, is the subject of much current research and new DNA sequencing techniques continue to improve our understanding of seaweed speciation. DNA-based molecular phylogenies are redrafting the lines drawn, sometimes incorrectly, between existing species. At the same time, we are uncovering hidden genetic variation, in which populations that look the same may be slowly moving apart.

This book lists more than 65 seaweed genera across three distinct phyla, but that only represents a small snapshot of Earth's changing seaweeds, and the seaweeds in the snapshot will not be with us forever: new species and new genera will arise to take their place. So, enjoy the ones we have while they are still with us!

LEFT | The green seaweeds *Ulva lactuca* (top) and *Monostroma grevillei* (bottom). Cells from *Ulva* and *Monostroma* may be artificially fused in the lab to form hybrid seaweeds, despite the fact that these species diverged from each other perhaps 170 MYA. Understanding how and why nature separated these species is the study of speciation.

In previous sections of this book, we have discussed what seaweeds are and why they are important. This section takes a closer look at their diversity. Living things are classified using the same basic method that Linnaeus proposed in the eighteenth century (see page 24). Every species is sorted into a hierarchical tree that reflects its evolutionary descent. The most familiar categories in this hierarchy, from largest to smallest, are:

Kingdom > Phylum > Class > Order > Family > Genus > Species

When a new seaweed species is discovered, it is classified according to a range of characteristics: cellular ultrastructure, methods of reproduction, and, increasingly, particular DNA sequences known as barcodes (because they help to identify things).

There are two things to bear in mind about seaweed taxonomy. First, life flows constantly and subtly. The distinct categories in the Linnaean hierarchy are useful tools, but rather blunt ones to describe the delicate shades with which evolution paints. There is often, therefore, taxonomic disagreement about exactly how different seaweeds need to be, to be sorted into different orders, and so on. Second, seaweeds are the products of endosymbiosis, which means that the branching Linnaean tree of life should more properly be thought of as a more interwoven web of life. The implications of this are still being explored.

SEAWEED DIVERSITY

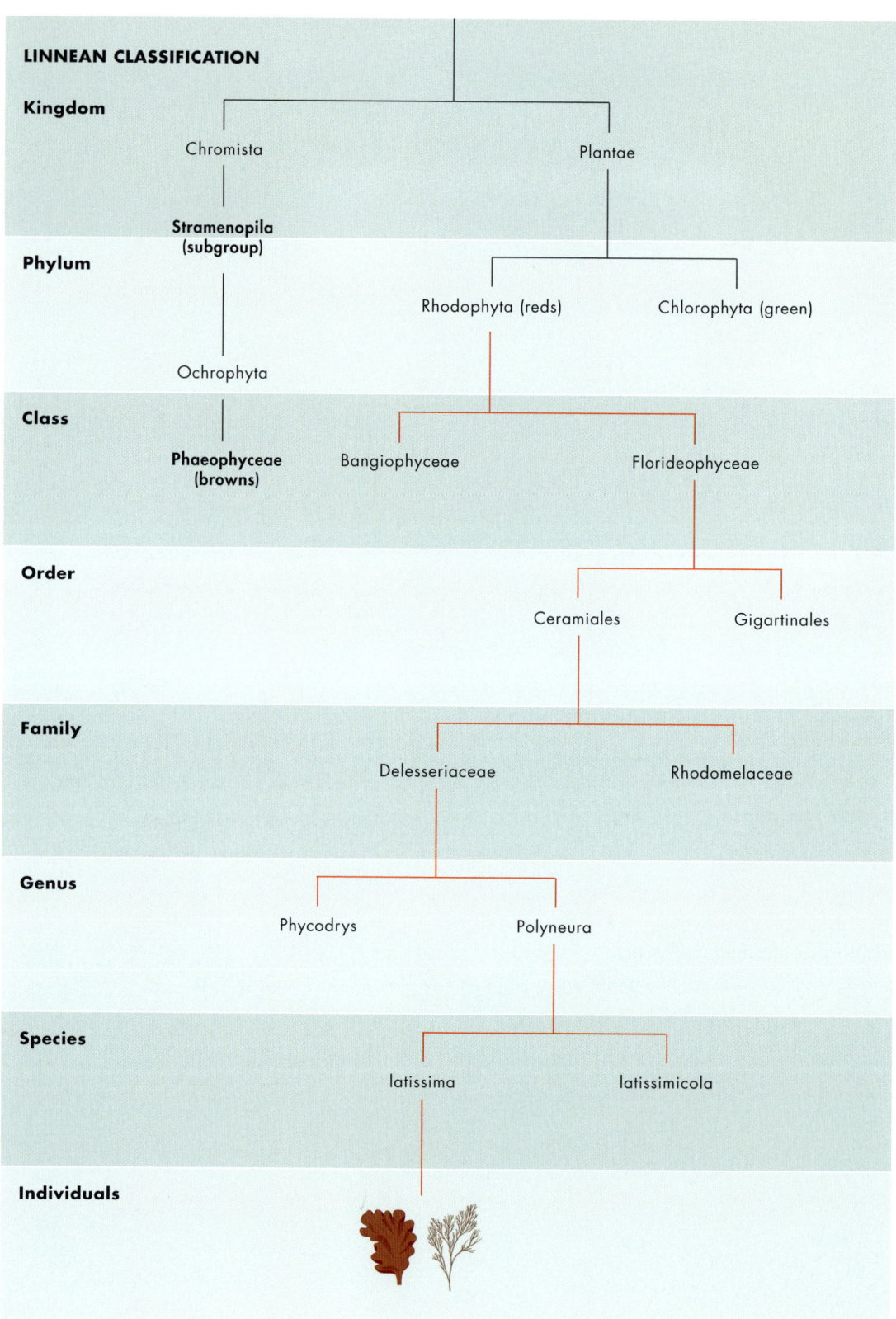

LINNEAN CLASSIFICATION

Kingdom

Chromista

Plantae

Phylum

Stramenopila (subgroup)

Rhodophyta (reds)

Chlorophyta (green)

Ochrophyta

Class

Phaeophyceae (browns)

Bangiophyceae

Florideophyceae

Order

Ceramiales

Gigartinales

Family

Delesseriaceae

Rhodomelaceae

Genus

Phycodrys

Polyneura

Species

latissima

latissimicola

Individuals

Kingdom: Members of the same kingdom share their fundamental cellular architecture, but often not much more than that. The red and green seaweeds share the kingdom Plantae, but the browns are over in the kingdom Chromista. This means that there is almost as much difference between the green *Ulva* and the brown *Fucus* as there is between a human and a potato.

Phylum: In seaweeds, these end in "-phyta." The green seaweeds all share the phylum Chlorophyta, the red seaweeds share the phylum Rhodophyta, and the brown seaweeds share the phylum Ochrophyta. We share our phylum, the Chordata, with the fish, the amphibians, and the sea squirts (we all have backbone-like structures). The red seaweeds *Palmaria* and *Porphyra* both look like flat red blades, but they come from different classes (Bangiophyceae and Florideophyceae) in the same phylum (Rhodophyta). We should expect there to be as many differences between them as there are between whales and fish.

Class: In seaweeds, all classes end in "-phyceae": Bangiophyceae, Florideophyceae, and so on. Species that share a class usually share fundamental aspects of their growth or development. Horses and humans are both members of the mammal class, for example, and we both feed our young on mothers' milk.

Order: Seaweed orders all end in "-ales": Bangiales, Ceramiales, and so on. Species that share an order usually share the same basic body architecture but can vary widely in their shapes and sizes. Horses share their order, Perissodactyla, with rhinoceroses. Primates are another order, as are rodents. Wracks and kelps are different orders of brown algae, so we should expect them to be as different from each other as humans are from mice.

Family: In the seaweeds, family names all end in "-aceae": Fucaceae, Ulvaceae, and so on. Individuals that share a family generally have the same overall body shape but you would not confuse them for each other. Humans are in the great ape family (Hominidae) along with gorillas and chimpanzees: we are all bipedal and have no tails, but gorillas are easy to distinguish from chimpanzees.

Genus: As a rule of thumb, if two species look similar enough that you sometimes confuse them, then they are probably in the same genus. They may even confuse each other, as species in the same genus can sometimes breed to produce hybrids. For example, horses (*Equus ferus*) share the genus *Equus* with donkeys (*Equus africanus*) and the two can mate to produce either a mule (if the donkey was the father) or a hinny (if the donkey was the mother). *Fucus vesiculosus* and *Fucus spiralis* are seaweed species in the same genus and hybridize in a similar fashion.

Species: A simple definition of a species is to say that it is a group of organisms that, between them, can reproduce themselves. If a male and a female can produce male and female offspring that can, in turn, produce male and female offspring, then the original male and female are members of the same species.

OPPOSITE | The full Linnean classification of the red seaweed *Polyneura latissima*. Every species has its variation on this.

SELECTED SPECIES

There are thousands of known seaweed species and well over 200 known genera, spread across three distinct phyla. In this book, we have included 75 genera, for the following reasons:

1. They were among those first described by Linnaeus in either his 1753 *Species Plantarum* or his 1759 *Systema Naturae*. Linnaeus chose those species because they were common and distinctive. Greens: *Cladophora, Spongomorpha, Ulva*. Reds: *Acanthophora, Botryocladia, Ceramium, Furcellaria, Gigartina, Griffithsia, Osmundaria, Palmaria, Phycodrys, Plocamium, Porphyra*. Browns: *Ascophyllum, Chorda, Cystoseira, Ericaria, Fucus, Halidrys, Halopteris, Himanthalia, Padina, Pelvetia, Pylaiella, Saccharina, Sargassum, Gongolaria, Turbinaria*.

2. Since most of Linnaeus' species were found in European waters, this selection also includes a number of genera that have significant cultural or economic importance worldwide. Greens: *Caulerpa, Codium, Monostroma*. Reds: *Ahnfeltia, Asparagopsis, Chondrus, Eucheuma, Gelidium, Gloiopeltis, Gracilaria, Grateloupia, Jania, Kappaphycus, Nemalion, Phymatolithon, Pyropia, Sarcopeltis*. Browns: *Analipus, Desmarestia, Dictyopteris, Durvillaea, Ecklonia, Egregia, Eisenia, Himantothallus, Laminaria* (pictured), *Macrocystis, Nemacystus, Nereocystis, Pelagophycus, Saccorhiza, Undaria*.

3. Some genera are included for their scientific interest or particular beauty. Greens: *Acetabularia, Halimeda, Valonia*. Reds: *Bostrychia, Ceratodictyon, Halosaccion, Hypnea, Polyneura, Pterothamnion*. Browns: *Cutleria, Ectocarpus* (included with *Pylaiella*), *Hormosira, Postelsia*.

TAXONOMIC TERMINOLOGY

Phycologists use a lot of jargon. While keeping it to a minimum, several scientific terms are used in the following pages. One thing to watch out for is that the green, red, and brown seaweeds often have their own terminology. The word "siphon," for example, is the Greek word for a tube and is often used differently in green and red seaweeds. In the greens, "siphonous" usually refers to the structure of cells. In the reds, it usually refers to the structure of the thallus. This is explained in more detail below.

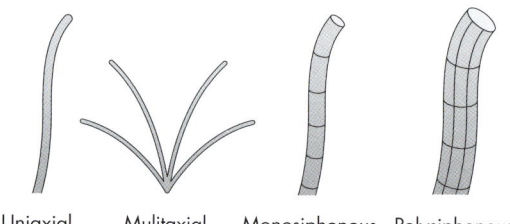

Uniaxial   Mulitaxial   Monosiphonous   Polysiphonous

ARRANGEMENT OF FRONDS IN A THALLUS

A seaweed may have one main axis (uniaxial) or several main axes (multiaxial). In the red seaweeds, each axis may consist of a single chain of cells (monosiphonous) or of several chains aligned to form a bundle (polysiphonous).

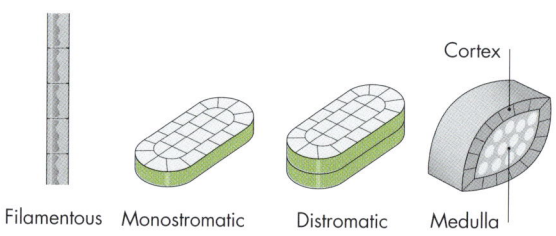

Filamentous   Monostromatic   Distromatic   Medulla   Cortex

ARRANGEMENT OF CELLS IN FRONDS

Many seaweeds consist of cells in a row: we call these filamentous. If the thallus consists of one cell layer, we call it monostromatic. If it contains two cell layers, one above the other, we call it distromatic. If the thallus is thicker and contains several layers, we call the outer layer the cortex (Latin for "outer shell") and the inner part the medulla (Latin for "marrow").

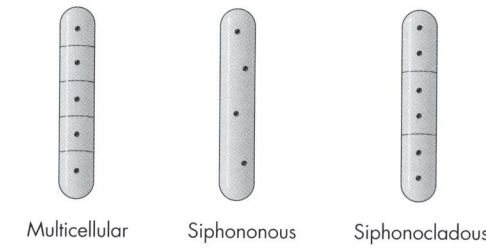

Multicellular   Siphononous   Siphonocladous

SEAWEED CELL STRUCTURE

Generally speaking, red and brown seaweeds are multicellular. Their thalli consist of lots of distinct cells and each cell has its own nucleus. The greens show more variety and come in three kinds: multicellular, siphonous, and siphonocladous. Some greens consist of one extended cell that contains many nuclei: these are called siphonous. If the cells are distinct, but each cell has several nuclei, then we call it siphonocladous.

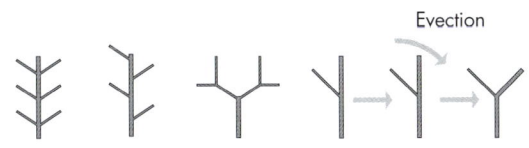

Opposite | Alternate | Dichotomous | Pseudo-dichotomous

Evection

BRANCHING PATTERNS

Seaweeds can display a number of branching patterns. Some common ones are shown above.

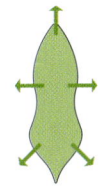  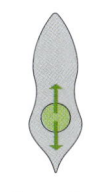

Diffuse | Apical | Intercalary

TYPES OF MERISTEMS

The part of the seaweed that grows is called the meristem. Meristems may be diffuse (that is, the whole seaweed grows), apical, or intercalary (Latin for "inserted into something").

Stolon | Rhizoid | Disc | Haptera

THE STRUCTURE OF HOLDFASTS

A simple horizontal tubular holdfast is called a rhizoid (Greek *rhizos* = "root"). In some genera, notably *Caulerpa*, rhizoids extend into a creeping horizontal tube from which fronds arise at regular intervals. This is called a stolon. Some holdfasts are disk-shaped, or discoid. Others are made of tangled tubes called haptera (from the Greek for "a fastening").

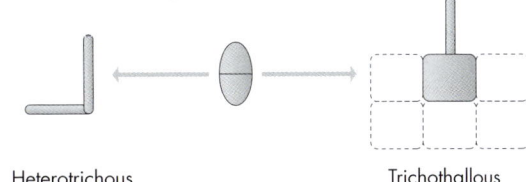

Heterotrichous | Trichothallous

PATTERNS OF GROWTH AT THE MERISTEM

There are two notable patterns of cell division that are seen especially in brown seaweeds. In heterotrichous growth (Greek for "different hair"), a cell divides to give one horizontal cell that becomes a tubular holdfast and one erect cell that grows into a frond. In trichothallic growth (Greek for "hair" + "green shoot") a cell divides to give an upper hair-like projection and a lower thallus cell.

Epiphyte
Growing on another seaweed but carrying out its own photosynthesis

Parasite
Growing on another seaweed and stealing its host's photosynthetic sugars

SEAWEEDS THAT GROW ON SEAWEEDS

Many seaweeds grow on or in other seaweeds. If a seaweed is anchored onto another living thing, but is otherwise free-living, we call it an epiphyte. If the seaweed is free-living but growing between the cell layers of another organism, we call it an endophyte. If the seaweed is both growing on and feeding on another living thing, it is a parasite.

Class

Order

Family

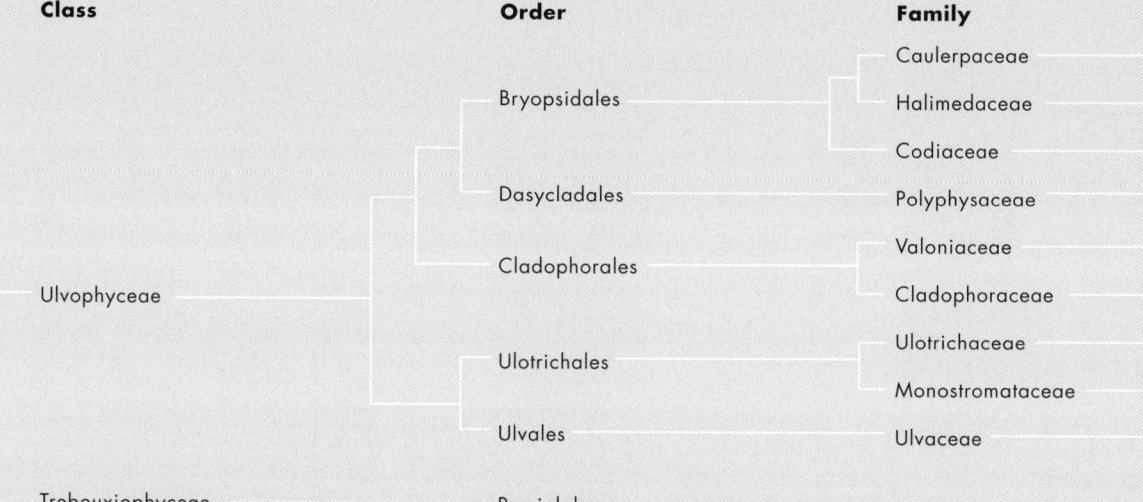

Ulvophyceae

Trebouxiophyceae

Bryopsidales

Dasycladales

Cladophorales

Ulotrichales

Ulvales

Prasiolales

Caulerpaceae

Halimedaceae

Codiaceae

Polyphysaceae

Valoniaceae

Cladophoraceae

Ulotrichaceae

Monostromataceae

Ulvaceae

LEFT | A probable phylogeny of the green seaweeds. As with all algae, the exact taxonomic relationships are the subject of much research, so this serves as a good current model.

CHLOROPHYTA
THE GREEN SEAWEEDS

Genus

Caulerpa

Halimeda

Codium

Acetabularia

Valonia

Cladophora

Spongomorpha

Monostroma

Ulva

Dominating the upper intertidal zone, the green seaweeds comprise fewer species than either the red or the brown seaweeds and are the multicellular marine members of a much larger evolutionary group called the Chlorophyta.

The green seaweeds diverged from their red algal cousins around 1.5 GYA and from the ancestors of land plants perhaps 1 GYA. They are, therefore, more closely related to land plants than they are to the other seaweeds. This shows in their fundamental biology: they use the same storage polymer, starch, as the land plants and the same photosynthetic pigments. The green seaweeds do have important differences to land plants, of course, which reflect their saltwater habitats. Green seaweed chloroplasts have pyrenoids to supply dissolved carbon dioxide to rubisco and their cell walls contain characteristic algal-specific soluble polymers (for example, ulvan rather than pectin). Their cell walls are structurally distinct from land plant cell walls and contain much less cellulose, with some green seaweed orders lacking cellulose altogether.

We currently think that green seaweeds are found in two classes of the Chlorophyta, the Ulvophyceae and the Trebouxiophyceae, which diverged from each other perhaps 800–850 MYA. Most are in the former class: there are at least five main green seaweed orders in the Ulvophyceae and only one in the Trebouxiophyceae. The Trebouxiophyceae are largely unicellular but one order, the Prasiolales, forms multicellular, bladed, seaweeds. The Ulvophyceae, on the other hand, contains five major seaweed orders: Bryopsidales, Cladophorales, Dasycladales, Ulvales, and Ulotrichales. These are thought to have diverged from each other perhaps 600–800 MYA.

CAULERPA

Like all of the Bryopsidales, *Caulerpa* are coenocytes (Greek *koinos* = "common" + *kytos* = "a cell"): rather than being divided into many microscopic cells, each with its own nucleus, the thallus of *Caulerpa* consists of one continuous cell that contains many nuclei. Phycologists call this siphonous (see page 78). *Caulerpa* are attractive seaweeds, growing from a horizontal tube called a stolon that creeps along a surface and, at regular intervals, sends up a leaflike frond. These fronds form a range of delicate species-specific shapes. You might think that it would be hard to make a fluid-filled bag stand up in the way that *Caulerpa* leaves do, and you would be right. To overcome this, *Caulerpa* have evolved a unique innovation called trabeculae (Latin for "little beam"), which are ingrowths of their cell walls that provide mechanical support as internal buttresses.

RIGHT | "Green caviar," one of the common names for *Caulerpa lentillifera* (Latin: "bead-bearing").

OPPOSITE | The tropical green seaweed *Caulerpa taxifolia* growing where it should not, off the south coast of France. Thought to have escaped from Monaco's Oceanographic Museum, the species is now invasive in the Mediterranean.

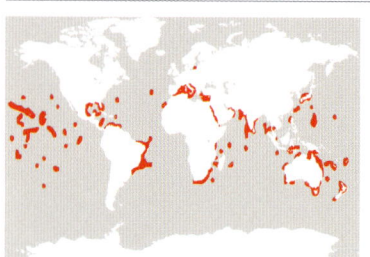

ETYMOLOGY
Greek *kaulos* = "stem" + *herpo* = "to creep"

DIVERSITY
Around 100 accepted species

SHAPE & SIZE
Bright green and leaflike, usually 4–40 in (10–100 cm) long

HABITAT
Upper intertidal zone in warm and temperate waters

LIFE CYCLE
Often reproduces by vegetative (asexual) propagation. The sexual life cycle remains mysterious but is likely based on a single diploid generation (the visible fronds, which may be monoecious or dioecious) that produces haploid gametes (see page 33)

Caulerpa species are common and important worldwide, with their horizontal stolons helping to stabilize sandy sediment coasts, especially around lagoons and reefs. Their slightly spicy taste makes them popular as "green caviar" and they are eaten as the *umibudo* of Japan and as a relish across many Pacific nations. As with most things, *Caulerpa* should be eaten in moderation because some species produce a chemical mixture called caulerpicin, which can be toxic at high concentrations. It is also rather unfortunate that some *Caulerpa* species can be *too* attractive: the delicate and fast-growing leaves of *Caulerpa taxifolia* have made it a popular saltwater aquarium plant and, as aquarium plants do, it can sometimes escape. As a result, large parts of the Mediterranean have seen serious and damaging invasions of *Caulerpa* and, to prevent similar ecosystem damage, it has been illegal to possess *C. taxifolia* in California since 2001.

CODIUM

The forest-green members of the *Codium* genus can adopt a wide range of forms, from the delicate fingers of *Codium fragile* to the distinctive squashed sac of *Codium bursa*, which looks like nothing so much as a burst green beach ball. One thing that they all have in common, however, is their sponginess. This makes them attractively edible and they are iodine and vitamin A rich. *Codium* species are eaten as the sea moss of Barbados, the *miru* of Japan, the *pokpoklo* of the Philippines, and the mousefoot *limu* of Hawaii. In Japan, for example, *C. fragile* is collected in spring, dried, and then preserved in ashes or salt.

BELOW | "Dead man's fingers."
Codium fragile (Latin: "delicate")
at Land's End, Cornwall, UK.

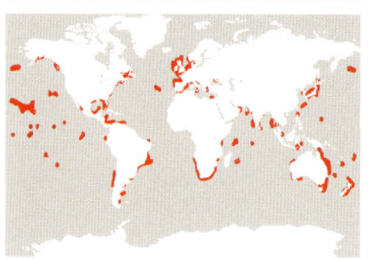

ETYMOLOGY
Greek *kodion* = "pelt" or "hide"

DIVERSITY
Around 150 accepted species

SHAPE & SIZE
Usually fairly small, reaching maybe 10 in (25 cm) across. They tend to be a deeper, more foresty, green than most chlorophytes

HABITAT
Upper intertidal zone worldwide

LIFE CYCLE
The visible fronds are the diploid adults, which can be male or female. These produce haploid gametes that regenerate the visible fronds (see page 33)

The sponginess and variety of *Codium* forms reflects their remarkable construction: *Codium* seaweeds are made up of thousands of microscopic tubular filaments that are woven together to form a multiaxial structure, in much the same way that straws may be woven together to form a corn dolly. As with all the Bryopsidales, these tubular filaments form a siphonous coenocyte: an interconnected fluid-filled network that contains thousands of nuclei. Even more impressively, when the filaments are near the surface of the *Codium* thallus, they pinch out a bleb that lengthens and expands into a rod-shaped structure called a utricle. These utricles then pack side by side, like logs in a stack of wood, to form the surface layer of the *Codium* thallus. When *Codium* becomes fertile, the reproductive gametangia grow out from the utricles as spurs.

ABOVE | The growths that look like squashed footballs are "sponge ball" seaweed, or *Codium bursa* (Latin: "pouch"), seen here in the Kornati archipelago off the coast of Croatia.

RIGHT | A microscope image of utricles—specially expanded sections of the cells of seaweeds in the Bryopsidales order. The surfaces of these seaweeds are made by stacking utricles next to each other.

ULVOPHYCEAE: BRYOPSIDALES: HALIMEDACEAE

HALIMEDA

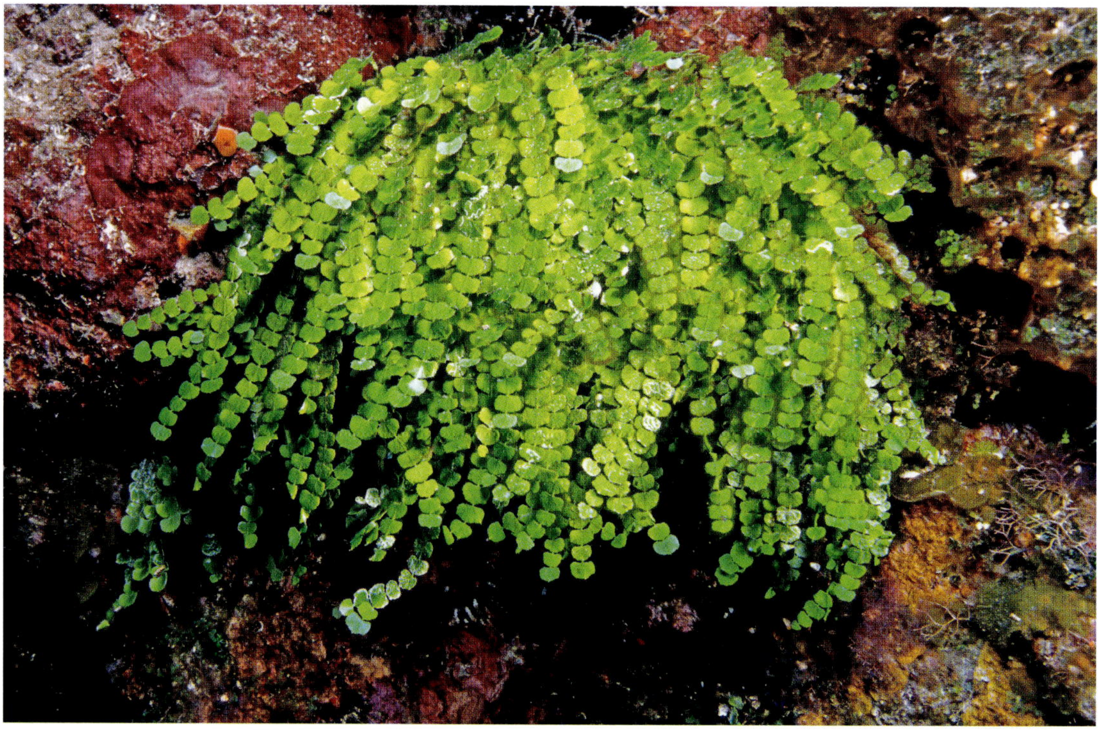

This is the first of several calcifying seaweeds to feature in the book (the others are the red maerls, *Phymatolithon* and *Jania*, and the brown *Padina*; see pages 110, 112, and 116). As with all Bryopsidales, the thalli of *Halimeda* are constructed from siphonous filaments. These weave together to form an underlayer (medulla) and rod-shaped outgrowths called utricles extend from this filamentous underlayer, packing tightly next to each other to form the surface (cortex) of the thallus. The thalli have holdfasts adapted to attach to rocks or sand, depending on the species.

Unusually, *Halimeda* is able to deposit crystals of calcium carbonate ($CaCO_3$) in the spaces between its

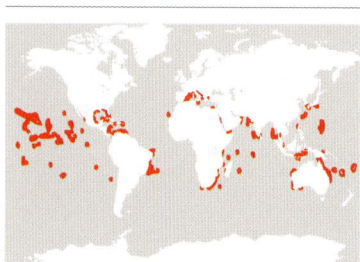

ETYMOLOGY
Greek *halimedon* = "guardian of the sea"

DIVERSITY
Around 50 accepted species, although cryptic species may bring this number up

SHAPE & SIZE
Distinct, coin-sized plates, joined in multiaxial clusters around 4–8 in (10–20 cm) long

HABITAT
Tropical

LIFE CYCLE
The visible fronds are the diploid adults, which can be male or female. These produce haploid gametes that regenerate the visible fronds (see page 33)

siphonous filaments, creating stiffened disks that are joined to each other by flexible, noncalcified hinges. Calcium carbonate can crystallize into three main geometries, represented by its three mineral forms: calcite, aragonite, and vaterite. The form seen in the thalli of *Halimeda* is aragonite, which has a more rectangular (orthorhombic) crystal lattice, and these aragonite crystals can end up contributing around three-quarters of the weight of the seaweed.

This calcification is of huge ecological importance. *Halimeda* is a warm-water species that grows in sheltered waters around coral reefs, where it undergoes what is called holocarpic reproduction. When thalli become fertile, the entire siphonous cell is converted into gametes and the release of these new gametes destroys the old thallus. This leaves the

OPPOSITE | *Halimeda goreaui*, named after German coral biologist Thomas Goreau (1924–70). Flattened disks are characteristic of *Halimeda* species.

ABOVE | *Halimeda discoidea* (Latin: "disks"), growing off Cassini Island, Western Australia.

now empty calcified skeleton, which disintegrates into a fine calcium carbonate sand. This sand can form an astonishing amount of the sediment that builds up coral atolls. Funafuti atoll, for example, sits around 2,000 miles (3,300 km) north of New Zealand over to the western edge of Polynesia. It is the capital of Tuvalu and we think that *Halimeda* sand forms around 80 percent of the top 65 ft (20 m) of the atoll.

ACETABULARIA

Members of the *Acetabularia* genus form beautiful, goblet-like structures in which the shape of the top section, or cap, varies from species to species. This distinctive shape gives rise to the common names of "mermaid's wineglass" or "mermaid's parasol" and, ultimately, to the name of the genus: an *acetabulum* was a roman cup that was used to hold a vinegar dip. *Acetabularia* species are small and relatively rare, being found underwater in subtropical areas, and they grow by an unusual process called diaphysis: new tissue forms inside old tissue and then pushes its way out, leaving a scar.

The graceful *Acetabularia* thallus is actually a single giant cell that is several inches long, with a single nucleus at the bottom of its stalk. This is astonishing, because most single-nucleus cells are smaller than the ~100 µm limit of the naked eye. Adult *Acetabularia* are diploid and usually twiglike and their caps are the reproductive structures, appearing when *Acetabularia* becomes fertile. When fertile, the nucleus at the bottom of the stalk divides by meiosis to give haploid nuclei and these replicate themselves to give thousands of copies, which stream up the stalk and into the new cap. Each haploid nucleus then becomes surrounded by a membrane

to form a small pouch, or cyst, and these cysts dock with the inside wall of the cap before opening a small hatch (or operculum) to the outside that allows the haploid nuclei to leave and float away as gametes.

For a species that is perhaps the hardest to find of any in this book, *Acetabularia* has made a disproportionate contribution to our understanding of biology. In the early part of the twentieth century, the German biologist Joachim Hämmerling realized that the cap of one *Acetabularia* species could be grafted to grow on the stalk of another. When Hämmerling transplanted the cap of *A. crenulata*

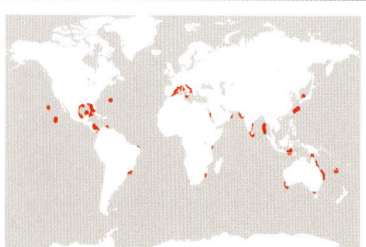

ETYMOLOGY
Latin *acetabulum* = "vinegar cup"

DIVERSITY
Around 20 accepted species

SHAPE & SIZE
Delicate, radially symmetric, cuplike structures that can reach a few inches (~10 cm) in length

HABITAT
Subtropical waters, including the Mediterranean Sea. Often grow several feet underwater in the subtidal zone

LIFE CYCLE
The thallus is the diploid adult and the gametes are haploid (see page 33)

onto *A. acetabulum*, he found that the *crenulata* cap adopted the shape of the host *acetabulum*. He inferred that the host *acetabulum* nucleus must be sending new instructions to the transplanted *crenulata* cap, making this one of the first demonstrations to show that the nucleus controlled the development of an organism.

VALONIA

Y ou are sadly unlikely to see *Valonia* species in the wild, unless you go diving in tropical waters; some species find their way into aquaria. Its most famous representative, *Valonia ventricosa*, is quite extraordinary, resembling a large and delicate green glass marble, anchored at the bottom by rhizoids. Its common name, not surprisingly, is "bubble algae." The structure of these bubbles is very unusual: they are sometimes given as examples of extremely large single cells but, as ever, things are slightly more complicated: at the center of a *Valonia* "bubble" sits a large central vacuole that takes up most of the bubble. Thousands of nuclei are then arranged around this central vacuole and each nucleus sits in its own domain of cytoplasm. An easy way to think about this is a strawberry: the seeds of the strawberry are arranged around the fruit in much the same way that the nuclei of *Valonia* are arranged around the central vacuole. The nuclei are much too small to see with the naked eye, however, so what we see is the cell wall stretched tightly over the vacuole. While *Valonia* species are thought to reproduce sexually, they can also propagate through a process called segregative

cell division. During segregative cell division, each cytoplasmic domain develops its own membrane and the bubble opens up to release these individual smaller cells.

ABOVE | "Bubble algae." The distinctive spheres of *Valonia ventricosa* (Latin: "belly-like"), framed by another green seaweed, *Halimeda*.

OPPOSITE | Spheres of *Valonia ventricosa* can be 4 in (10 cm) wide. This one is covered in debris, but the glossy cell membrane shines through.

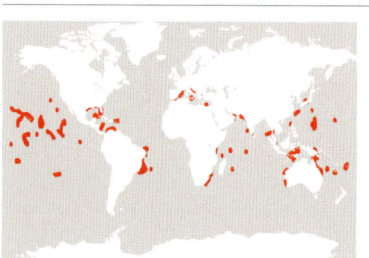

ETYMOLOGY
Italian *valonia* = "acorns"

DIVERSITY
Around 20 accepted species

SHAPE & SIZE
A distinctive, marble-like appearance, with some species forming colonies that look like grapes; usually a few inches across but the largest can reach lengths of around 8 in (20 cm)

HABITAT
Tropical and subtropical areas. Often found relatively deep (30–100 ft/10–30 m) in the subtidal zone

LIFE CYCLE
Thought to have sexual isomorphic alternation of generations (see page 33), although vegetative propagation by fragmentation of the thallus may be more common

CLADOPHORA

The Cladophorales are siphonocladous, which means that each thallus is made up of many siphonous cells and each of those cells contains many nuclei (see page 78). These nuclei can replicate without dividing to become polyploid and, when they do, new *Cladophora* species can arise. Humans, obviously, are not used to this kind of arrangement, because humans (and most animals) have multicellular bodies in which each cell has only one diploid nucleus. *Cladophora* cells are strung together in chains to form filaments and these filaments can branch, so *Cladophora* species often look tufty and mosslike. The genus is common worldwide and its species are often eaten.

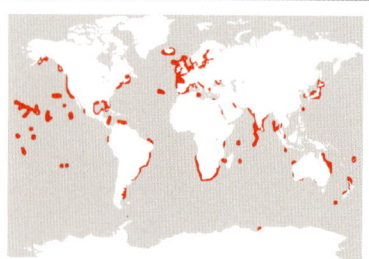

ETYMOLOGY
Greek *klados* = "branch" + *phora* = "bearer"

DIVERSITY
Around 300 accepted species

SHAPE & SIZE
Filamentous and branched; usually a vivid green and small (an inch or so in length)

HABITAT
Fresh or brackish water worldwide, so may be found in estuaries and the upper intertidal zone

LIFE CYCLE
Isomorphic alternation of generations (see page 33)

They are, for example, pounded and dried into the *kaipen* of Laos. *Cladophora* may be male or female (dioecious) and the male and female gametes they produce look the same as each other (isogamous).

For many people, the name *Cladophora* will be familiar from *Cladophora* balls, or the *marimo* of Japan (Japanese *mari* = "ball" + *mo* = "water plant"). *Marimo* are about the size and shape of a beach ball and are found in colder freshwater lakes and rivers in Japan and Europe. *Cladophora* balls are sometimes called moss balls but they are, in fact, neither *Cladophora* nor moss. Instead, they are a rare thallus form of yet another macroalga called *Aegagropila linnaei*. *Cladophora* and *Aegagropila* are similar enough that *Aegagropila* used to be classified as *Cladophora* (hence "Cladophora ball") but *A. linnaei* is freshwater, rather than marine, so is not a true "sea" weed. *Cladophora* and *Aegagropila* are both members of the same order, Cladophorales, and *marimo* are woven from siphonous filaments in just the same way as members of *Cladophora*. As a final twist in this story, the formation of these algal balls from siphonocladous filaments is thought to be helped by rolling and revolving currents. These occur more often in lakes, but are sometimes found along the coast and, under these conditions, true seaweed "beach balls" (usually Cladophorales) are sometimes found on beaches around the world.

OPPOSITE | *Cladopora rupestris* (Latin: "rock-dwelling") growing in long tangled mats.

ABOVE | A close relative of *Cladophora*, *Aegagropila linnaei*, showing how the long filaments of the Cladophorales can sometimes be rolled into balls.

RIGHT | Single-cell-wide filaments of *Cladophora* under the microscope.

SPONGOMORPHA

Species of *Spongomorpha* are distinctive enough to have been selected by Linnaeus as one of his original seaweed types. They are small and grow initially as flat bundles of filaments that weave together into a rough disk. Some of these filaments will begin to grow vertically to form the longer filaments of the thallus.

As their name suggests, all of these filaments are spongy and *Spongomorpha* are often eaten. Unusually, the visible *Spongomorpha* weed is actually the haploid gametophyte generation; the diploid sporophyte is microscopic and was originally misidentified as a separate species and given the name *Codiolum*. This microscopic "*Codiolum*" stage (the diploid sporophyte) grows endophytically on the thalli of other seaweeds and releases haploid zoospores that have four flagella and regenerate the visible filamentous seaweed (the haploid gametophyte generation).

ABOVE | The haploid gametophyte of *Spongomorpha aeruginosa* (Latin: "verdigris-colored"), growing here off County Clare, Ireland.

ETYMOLOGY
Greek *spongos* = "sponge" + *morphe* = "form"

DIVERSITY
Around ten accepted species

SHAPE & SIZE
Uninucleate branching filaments; larger species can be a few inches long

HABITAT
Lower intertidal zone worldwide and some species are epiphytic

LIFE CYCLE
Heteromorphic alternation of generations with the haploid gametophyte being much larger than the microscopic diploid sporophyte (see page 33); dioecious

MONOSTROMA

Cultivated as common foods along the coastlines of Japan and China, *Monostroma* species are among those that make up the Japanese food *aonori* and are particularly rich in vitamin B_1. Despite this economic importance, their biology is relatively understudied. As their name suggests, *Monostroma* species are monostromatic and formed from a single thin sheet of cells that develops out of an inverted cuplike germling. These delicate, membrane-like leaves are part of their appeal as a foodstuff. The cells in this single sheet divide and proliferate in a regular manner and, under the microscope, their arrangement can resemble that of the blocks in New York City's famous grid plan. As with their cousins in the *Ulva* genus, this simple and regular growth in *Monostroma* species can be remarkably plastic, with *Monostroma* species being able to grow as flat sheets or as cylinders. The reproduction of *Monostroma* species is also very variable: some are only known to reproduce asexually, while others have heteromorphic life cycles, with the visible thalli being haploid gametophytes that cycle through microscopic diploid zygotes.

ABOVE | The leafy green blades of *Monostroma grevillei*, named after the British botanist and abolitionist Robert Greville (1794–1866). The Christmas-treelike species on the right is a young brown seaweed, *Osmundaria hybrida*.

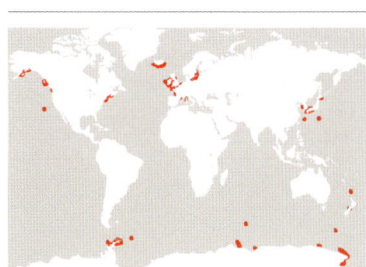

ETYMOLOGY
Greek *mono* = "one" + stroma = "layer"

DIVERSITY
Around 30 accepted species

SIZE
Variable, but some species can be 8–12 in (20–30 cm) in length

HABITAT
Monostroma species grow in fresh or brackish waters, so are found in the upper intertidal zone worldwide

LIFE CYCLE
Heteromorphic alternation of generations, with a very reduced diploid zygote (see page 33)

The *Ulva* genus is one of the easiest to find. Its members are cosmopolitan and common, growing to cover many upper intertidal shorelines with distinctive lawns of vivid green leaves. *Ulva* species can also contribute to nuisance blooms— the so-called green tides that are increasingly

ABOVE | A microscope image showing one layer (stroma) of cells in the *Ulva* thallus. In *Ulva*'s sea lettuce form, the thallus is simply two of these layers. In the tubular form, the space between the layers fills up with water to create a tube.

LEFT | One of *Ulva*'s two growth forms, the thin tubular form often called "threadweed" or "gutweed."

prevalent in the Yellow Sea, particularly along the shoreline north of Shanghai.

While the genus is easy to find, however, its species can be extremely hard to tell apart because their growth is so simple that two species will often end up looking very similar. *Ulva* thalli have very straightforward architectures, comprising two layers

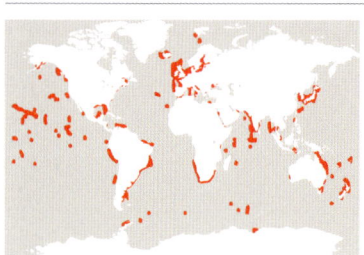

ETYMOLOGY
Latin *ulva* = "sedge" or "grass"

DIVERSITY
Around 100 accepted species

SHAPE & SIZE
Variable, but two major shapes exist: flat and leaflike or tubular and noodlelike. Most species are a vivid green and the larger species can be around 40 in (100 cm) in length

HABITAT
Upper intertidal zone; most species are attached, but some are free-floating; common worldwide

LIFE CYCLE
Isomorphic alternation of generations; many species are dioecious (see page 33)

ABOVE | The other growth form seen in *Ulva*. This is the leafy, bladed form that is often called "sea lettuce."

of cells that can either be flattened together to form leaflike blades or inflated apart to form tubular ribbons. This is reflected in the common names "sea lettuce" and "thread weed." The exact form adopted may depend on environmental conditions, so a single species can adopt either, or both, forms. To complicate things further, research shows that proper growth in at least some *Ulva* species requires a symbiosis with marine *Marinobacter* and *Roseovarius* bacteria. This impressive morphological flexibility makes *Ulva* species hard to identify unless one has a DNA sequencer to hand, but it also makes them a fascinating model for how marine species cooperate to build multispecies communities.

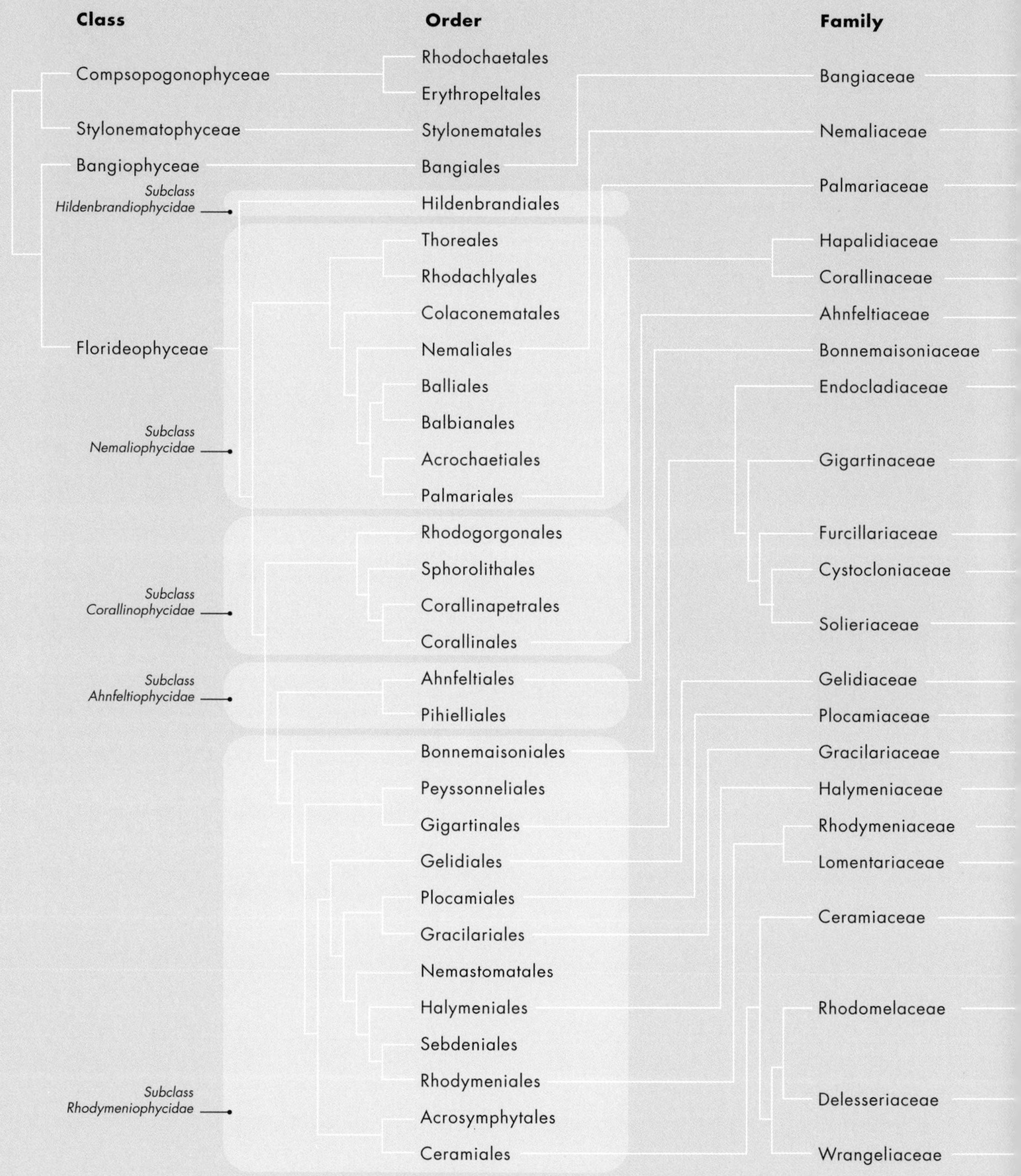

Class	Order	Family
Class	**Order**	**Family**

Compsopogonophyceae — Rhodochaetales

Erythropeltales — Bangiaceae

Stylonematophyceae — Stylonematales — Nemaliaceae

Bangiophyceae — Bangiales

Subclass Hildenbrandiophycidae — Hildenbrandiales — Palmariaceae

Thoreales — Hapalidiaceae

Rhodachlyales — Corallinaceae

Colaconematales — Ahnfeltiaceae

Florideophyceae — Nemaliales — Bonnemaisoniaceae

Balliales — Endocladiaceae

Balbianales

Subclass Nemaliophycidae — Acrochaetiales — Gigartinaceae

Palmariales

Rhodogorgonales — Furcillariaceae

Sphorolithales — Cystocloniaceae

Subclass Corallinophycidae — Corallinapetrales

Corallinales — Solieriaceae

Subclass Ahnfeltiophycidae — Ahnfeltiales — Gelidiaceae

Pihielliales — Plocamiaceae

Bonnemaisoniales — Gracilariaceae

Peyssonneliales — Halymeniaceae

Gigartinales — Rhodymeniaceae

Gelidiales — Lomentariaceae

Plocamiales

Gracilariales — Ceramiaceae

Nemastomatales

Halymeniales

Sebdeniales — Rhodomelaceae

Rhodymeniales

Subclass Rhodymeniophycidae — Acrosymphytales — Delesseriaceae

Ceramiales — Wrangeliaceae

LEFT | A possible phylogenetic tree of red seaweeds. Red algal phylogenies are often reassessed and species are not evenly distributed across this tree. Around 90 percent of known seaweed species are in the class Florideophyceae and around one third of known species are in one order, the Ceramiales. There are several red microalgal orders scattered across the reds, but those are not shown here.

RHODOPHYTA
THE RED SEAWEEDS

Red seaweeds are found worldwide, often dominating life in warmer subtidal waters. They are the crop species of the nori, carrageenan, and agar industries and contain the largest number of seaweed species—currently some 7,500 species across around 30 orders. As with the green seaweeds, the reds are the multicellular marine members of a large Phylum called the Rhodophyta, or red algae.

Red algae are characterized by three major evolutionary innovations. The first is their biochemistry. Red seaweed cell walls contain distinctive carbohydrate polymers that vary between species and include agar and carrageenan. Also, while the red and green seaweeds acquired their chloroplasts from the same primary endosymbiosis and so share the same basic photosynthetic pigments, the reds have evolved additional pigments that allow them to harvest more light, and so live farther from the surface.

Second, the red seaweeds have characteristic structures called pit connections, formed when red algal cells divide. Cell walls pinch inward, closing rather like a camera shutter to leave a small hole in the middle: this is the pit connection. Red seaweeds stop this hole up with structures called pit plugs, which have evolved through the red seaweed lineages. Third, red seaweeds are unusual because none of their cells have any flagella, which means that their reproductive cells cannot move very far.

Red seaweeds are currently divided among four major classes. Small and often epiphytic red seaweeds are found in the basal classes of the Compsopogonophyceae and Stylonematophyceae. The most familiar red seaweeds, however, are found in the subphylum Eurhodophytina, which contains the two classes of the Bangiophyceae and the Florideophyceae. The latter class is divided into five main subclasses: the Ahnfeltiophycidae, the Hildenbrandiophycidae, the Nemaliophycidae, the Corallinophycidae, and the Rhodymeniophycidae.

PORPHYRA

The leafy reds of the Bangiales are some of the most widely eaten seaweeds, with *Porphyra* providing the famous laver of Wales. Its rust-colored blades are high in protein and B vitamins, and are rich enough in vitamin C that these seaweeds were eaten to prevent scurvy on sailing ships in the eighteenth and nineteenth centuries. This long usage is matched by long study: *Porphyra umbilicalis* was first described by Linnaeus as *Ulva umbilicalis* in 1753, before being reassigned to the genus *Porphyra* by the German pharmacist and phycologist Friedrich Kützing in 1843.

Porphyra species are usually found in cooler waters, especially in the Atlantic. The blade is the haploid gametophyte and is thin, usually consisting of a single layer of cells (monostromatic). It can vary widely in shape, from circular to spearlike, and is linked to a disk-shaped holdfast by small, specialized rhizoid cells. The simple morphologies of both the haploid blade and diploid conchocelis generations are now being investigated using genetic clues from the *P. umbilicalis* genome, which was sequenced in 2017. This lacks many of the genes that usually build the frames of eukaryotic cells and shows a number of adaptations to life in the harsh intertidal zone, including genes that are thought to improve nutrient uptake and protect against desiccation and light stress.

The most famous research into *P. umbilicalis* was the elucidation of its life cycle by Kathleen Drew-Baker in the twentieth century, but much remains unknown about how *Porphyra* reproduces. For example, *P. umbilicalis* can reproduce sexually off European coastlines in the northeast Atlantic, but in

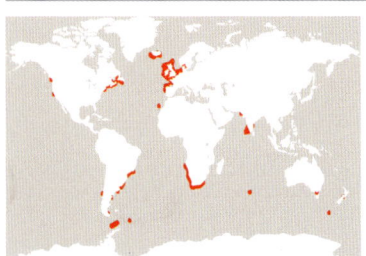

ETYMOLOGY
Greek *porphureos* = "purple"

DIVERSITY
Around 60 accepted species

SHAPE & SIZE
Leafy, reaching perhaps 8 in (20 cm) across

HABITAT
Intertidal zone worldwide, common in the North Atlantic

LIFE CYCLE
The blade is the haploid gametophyte, regenerating through a much smaller diploid conchocelis phase (see page 34)

the northwest Atlantic it only reproduces asexually, off the United States and Canada. We do not fully understand why, but different *Porphyra* species often intermingle, so the apparently asexual *P. umbilicalis* may be forming hybrids with other species such as *P. linearis*. This could, in time, give rise to new species.

BELOW | Laver. *Porphyra umbilicalis* (Latin: "dimpled") growing off the coast of Sark, one of the islands in the English Channel.

OPPOSITE | A blade of *Porphyra* at high tide on the south coast of Devon, UK.

RIGHT | A shell on which the conchocelis phase of *Porphyra* is growing. The visible dark dots are the fine filaments of the conchocelis generation.

PYROPIA AND NEOPYROPIA

LEFT | The leafy red blades of *Porphyra leucosticta*.

OPPOSITE | An early-nineteenth-century woodblock print by Utagawa Kuniyoshi (1797–1861), showing seaweed being harvested at Omori, to the south of Tokyo, Japan.

The broad, thin blades of *Pyropia*, with their small discoid holdfasts, look remarkably similar to the leafy blades of many *Porphyra* species, even though the two genera are thought to have diverged around 300 MYA. Much molecular work is being done on both genera and many species have been removed from *Porphyra* and reassigned into *Pyropia* and the closely related *Neopyropia*.

The best-known *Pyropia* and *Neopyropia* species are perhaps *P. columbina*, which is eaten as *karengo* by the Maori, and the many species that form part of the *nori* of Japan, the *zi-cai* of China, and the *gim* of Korea. Different species correspond to different varieties of these foods: *N. tenera* is amanori, for example, while *N. yezoensis* is susabinori.

Of these, *N. yezoensis* is the most widely cultivated, being grown across Japan, Korea, and China. In Edo period Japan (1603–1868), fishermen in Tokyo Bay and along the south coast would stake out bamboo bundles in fall and hope they were settled by juvenile *Pyropia* or *Neopyropia*. Their hard work was subject to the usual highs and lows of mariculture: annual harvests could deliver bumper crops or fail altogether. This changed in the mid-

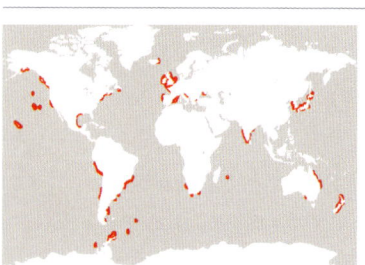

ETYMOLOGY
Greek *pur* = "fiery" + *op* = "face"; Greek *neo* = "new"

DIVERSITY
Around 70 accepted species

SHAPE & SIZE
Leafy, reaching perhaps 8 in (20 cm) across

HABITAT
Intertidal zone worldwide, including the Antarctic

LIFE CYCLE
The blade is the haploid gametophyte, regenerating through a much smaller diploid conchocelis phase (see page 34)

twentieth century with Drew-Baker's discovery that the various conchocelis-like seaweeds were, in fact, the tetrasporophyte phases of *Porphyra* and *Pyropia* species. Today, the conchocelis phase is maintained in tanks over the summer and seeded onto nets, rather than the traditional bamboo stakes, allowing for more reliable harvests: nets can give yields of around 4 lb per sq ft (20 kg per sq m). Pollution is an increasing problem and farmers are selectively breeding new cultivars of existing species to ensure a sustainable and resilient nori industry.

Pyropia and *Neopyropia* species are well adapted to harsh intertidal conditions, which is one of the reasons why they make good crops. They can survive losing around 90 percent of their water with each tidal cycle, recovering quickly once reimmersed.

P. endiviifolia has even adapted to the Antarctic. The genome of *N. yezoensis* was sequenced in 2013 and sequenced a little better in 2020, and much research is being done to understand the hardiness of these commercially important species.

BELOW | A plate of the Korean seaweed dish, *gim.*

NEMALION

The *umi-somen* of Japanese cuisine, *Nemalion* species are usually a darkish or brownish red and are distinctive, looking very much like the noodles that they are eaten as. *Nemalion* has multiaxial fronds, which means that multiple, comparably long fronds grow from one holdfast—this is sometimes called fountainlike growth. The fronds can branch and have rhizoidal (rootlike) holdfasts. In Japan, these slippery fronds are dried and preserved and eaten as a garnish on soup.

Nemalion gives its name to the subclass Nemaliophycidae, which it shares with the seaweeds on the next few pages (*Palmaria* and *Halosaccion*). The Nemaliophycidae are one of the earliest branching subclasses of the Florideophyceae and have their characteristic triphasic life cycle, although the carposporophytes in the Nemaliophycidae are formed directly from the carpogonium without the involvement of any auxiliary cells (see page 35).

The *Nemalion* generations are heteromorphic. The visible, slippery, noodlelike thallus is the haploid gametophyte, and the diploid tetrasporophyte is small and filamentous. The fronds of the haploid gametophyte thallus are solid, rather than fluid-filled, which is one of the reasons why they make good noodles. They also show some internal structure. The internal medullary cells are large and colorless;

ABOVE | The noodlelike fronds of *Nemalion elminthoides* (Greek: "wormlike") growing over barnacles and mussels.

ETYMOLOGY
Greek *nem* = "thread" + *hal* = "clump"

DIVERSITY
Around 10–20 species known

SHAPE & SIZE
Tangled noodles, with larger individuals being perhaps 20 in (50 cm) in length

HABITAT
Intertidal zone and brackish waters

LIFE CYCLE
Triphasic alternation of heteromorphic generations (see page 35); some species are monoecious and some dioecious

the outer cortical cells are smaller and pigmented and can produce filaments. When fertile, these filaments develop distinct spermatangia and carpogonia and there is some nuance. For example, some *Nemalion* species produce spermatangia first and carpogonia later, avoiding self-fertilization.

The filamentous diploid tetrasporophyte can reproduce sexually by making haploid tetraspores (which grow back into the noodlelike gametophyte) but can also propagate asexually by producing diploid monospores (which recycle the tetrasporophyte).

TOP | A seaside bouquet of *Nemalion elminthoides* wrapped in the green seaweed *Ulva*.

ABOVE | A cross section through a spongy *Nemalion* frond. The dark cortical cells on the outer surface and the lighter inner medullary cells are both clearly visible.

PALMARIA

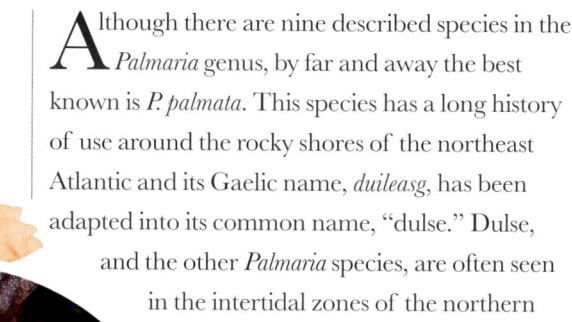

Although there are nine described species in the *Palmaria* genus, by far and away the best known is *P. palmata*. This species has a long history of use around the rocky shores of the northeast Atlantic and its Gaelic name, *duileasg*, has been adapted into its common name, "dulse." Dulse, and the other *Palmaria* species, are often seen in the intertidal zones of the northern hemisphere, where they can be found on rocks or as epiphytes on kelps. They can also support epiphytes themselves. They grow as a blade from a disk-shaped holdfast and in older individuals the blade can divide dichotomously to give a branching, fan-shaped thallus. The blade is reasonably fleshy, with an outer cortex and an inner medulla. This fleshy, leafy thallus made dulse a major source of nutrition for coastal communities before the industrialization of agriculture. Dulse was eaten particularly in Scotland, Iceland, and Ireland, with the practice spreading to the shores of Canada and the northeast United States as people migrated from Europe across the Atlantic. Men would head out to sea to fish while the "dulse-wives" waded out to collect

TOP | The ruby-red fronds of *Palmaria palmata* (Latin: "palm-like"; the species name repeats the genus name).

ABOVE | *Palmaria palmata* (to the right) growing alongside the brown seaweed *Fucus serratus*. The blades often look a deep red when out of water.

OPPOSITE | Dulse (*Palmaria palmata*) collected on the Northumberland coastline, UK.

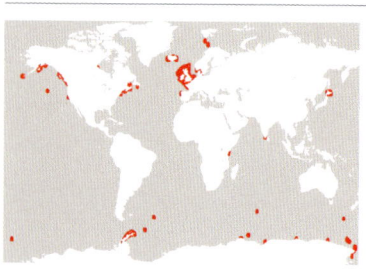

ETYMOLOGY
Latin *palma*, a loan word from the Greek *palam* = "hand"

DIVERSITY
Around ten accepted species

SHAPE & SIZE
Commonly found as rich purple, hand-shaped, dichotomous fronds or single blades, drying almost to black; can reach around 20 in (50 cm) long

HABITAT
Intertidal or subtidal zone; found in both hemispheres at temperate and polar latitudes, and flourishes in summer

LIFE CYCLE
Modified isomorphic (male gametophyte is isomorphic with tetrasporophyte; female gametophyte is different: tiny and microscopic). Unusually for the Florideophyceae, there is no carposporophyte phase (see page 35)

P. palmata by hand (see page 55). Dulse is rich in vitamins and minerals, particularly iodine, and is so nutritious that it was thought to have quasimedicinal properties. In 1806, the Scottish naturalist Patrick Neill wrote: "who eats of the dulse [. . .] will escape all maladies." Dulse is still widely eaten today, although as a luxury rather than a staple.

Palmaria also has an unusual life cycle. The genus gives its name to its order, the Palmariales, and these are unusual among the class Florideophyceae in having lost the carposporophyte phase of the complex triphasic life cycle. Instead, the haploid male spermatia fertilize the haploid female carpogonia and grow immediately into the diploid tetrasporophyte phase, with no intervening diploid carposporophyte phase. This life cycle reduction gives the Palmariales only two multicellular generations, rather than three, and is not seen in any other red orders. To add to the differences, the haploid gametophytes show severe sexual dimorphism. The males are large and leafy and look the same as (that is, they are isomorphic with) the diploid tetrasporophyte, but the female gametophytes are microscopic. They are fertilized by the haploid male spermatia and the resulting diploid tetrasporophyte grows out of the fertilized female carpogonium, quickly overshadowing it.

HALOSACCION

Members of the unmistakable Pacific *Halosaccion* genus have distinctive, saclike thalli, which explains its common name of "dead man's fingers." Its species are annuals and the sacs grow in clusters on rocky shores, being attached by short stipes to a disk-shaped holdfast. *Halosaccion* can look like dusty red grapes, but often appear as a yellowish-brown color that would be more suited to the brown seaweeds.

Their distinctive sacs allow *Halosaccion* species to cope with the constant drying out that is part of life on the shoreline. The thallus is pierced by microscopic pores. At high tide, when the thallus is underwater, the pores allow water in to fill the sac, creating something that looks a lot like a water-filled balloon. At low tide, when *Halosaccion* is out of the water (emersed) and drying, the pores slowly release that water, keeping the thallus moist.

The life cycle of one species, *H. glandiforme*, has been looked at in some detail and, like *Palmaria*, shows modified isomorphic alternation of generations without any carposporophyte phase. The sacs may either be diploid tetrasporophytes or haploid male gametophytes. The haploid female gametophytes are, instead, filamentous and microscopic. The diploid tetrasporophytes, which adopt the saclike thallus, produce haploid tetraspores

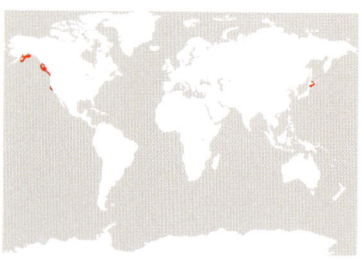

ETYMOLOGY
Greek *halo* = "sea" + *saccion* = "small bag"

DIVERSITY
Around ten species are known

SHAPE & SIZE
Sacs are usually around 2–4 in (5–10 cm) long

HABITAT
Shallow intertidal zone on both sides of the North Pacific; fairly tough and likes exposed rocks

LIFE CYCLE
Modified isomorphic (male gametophyte is isomorphic with tetrasporophyte; female gametophyte is different: tiny and microscopic); unusually for the Florideophyceae, there is no carposporophyte phase (see page 35)

in groups of four: two male and two female. The haploid female tetraspores develop rapidly into the small, fertile, filamentous female gametophytes, while the haploid male tetraspores grow more slowly into the male gametophyte, which again adopts *Halosaccion*'s characteristic saclike thallus. The male gametophytes overwinter and become fertile the following year, releasing spermatia in summer.

This complicated life cycle means that each year's female generation is fertilized by the previous year's male generation.

FLORIDEOPHYCEAE (CORALLINOPHYCIDAE): CORALLINALES: HAPALIDIACEAE

PHYMATOLITHON

The genus *Phymatolithon* is included as the first of two kinds of calcifying red seaweeds in the subclass Corallinophycidae and the order Corallinales. Unusually, genera in this order are able to deposit chalk in their cell walls. Chalk is the salt, calcium carbonate ($CaCO_3$), packed into a sloping (rhombohedral) crystal called calcite.

The crustose thalli of the Corallinales are common in warmer waters, where they help form maerl: beds of chalky red seaweeds. Traditionally, Breton farmers raked these maerl beds and ground the collected maerl into a chalky powder. This was then applied to farmland every couple of years to help fertilize acidic or peaty soils.

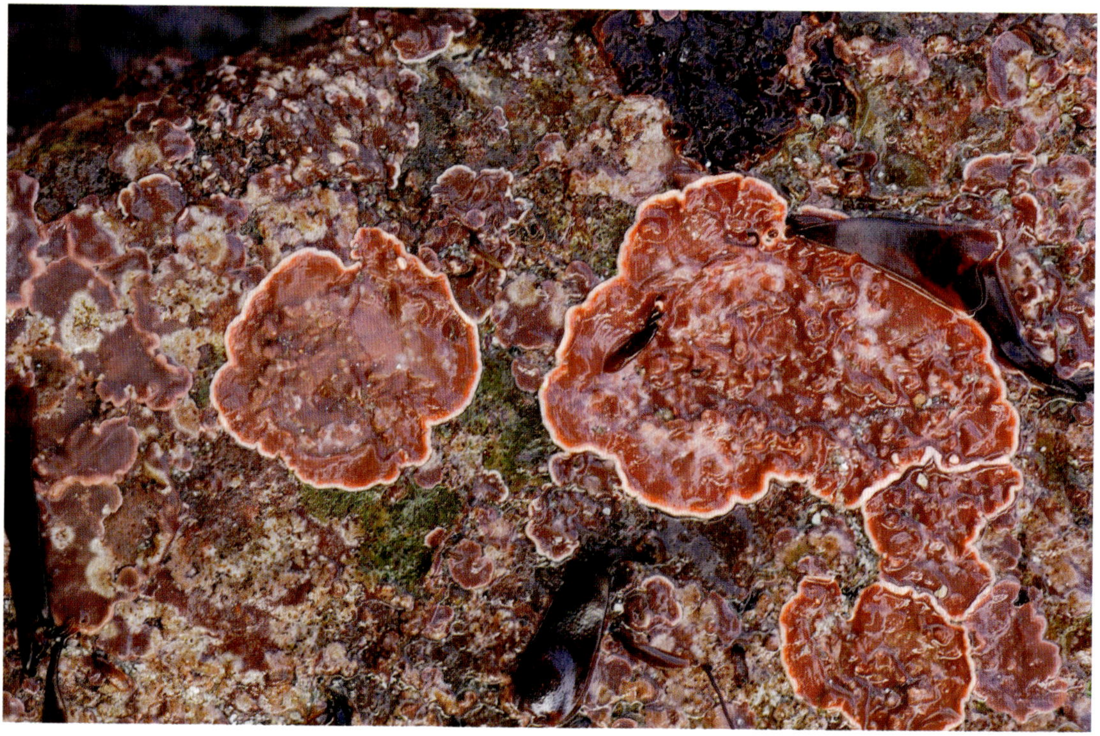

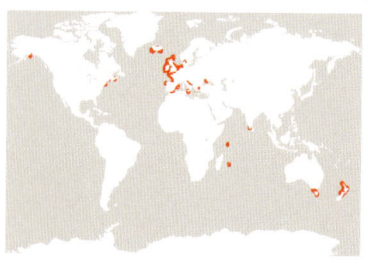

ETYMOLOGY
Greek *phyma* = "a growth" + *lithos* = "stone"

DIVERSITY
Around 20 accepted species

SHAPE & SIZE
A few inches across

HABITAT
Subtidal zone, often in maerl beds

LIFE CYCLE
Triphasic, with isomorphic gametophyte and tetrasporophyte (see page 35)

The first kind of calcifying red seaweed deposits chalk evenly into all of its cell walls to give the thallus a continuous chalk crust. The second kind leaves some gaps in the chalk to act as hinges. The two kinds are usually called non-geniculate and geniculate, from the Latin *geniculum*, meaning "knee," but are sometimes alternatively referred to as non-articulated and articulated.

Phymatolithon is one of the major non-geniculate genera and its sexual generations (the gametophyte and tetrasporophyte) grow as isomorphic crusts. Their chalk coating causes problems, not the least of which is ensuring that *Phymatolithon*'s reproductive cells can get out of the crust. To allow this, the reproductive conceptacles in fertile *Phymatolithon* develop tiny pores that pass through the chalk. In the haploid gametophyte, conceptacles have a single pore (uniporate); in the diploid tetrasporophyte, they have several pores (multiporate), although a microscope is needed to see the difference.

A second problem is that most florideophytes grow by generating new cells from apical meristems and then allowing those new cells to elongate. Chalk armor complicates that elongation. To get around this, *Phymatolithon* species have, unusually among the red algae, evolved a second meristem. The apical meristem generates the usual layer of new cells, which join the bottom of the crust (the hypothallus), but it also generates a second layer of cells that spread across the roof of the crust. This second layer is itself an intercalary meristem, producing new layers above (epithallus) and below itself (perithallus) with the epithallus being able to slough off to protect the rest of the crust.

Sexual reproduction in *Phymatolithon* seems to be limited to warmer waters. In the colder waters around the United Kingdom and Ireland, *Phymatolithon* reproduces by vegetative propagation: the tetrasporophyte has adopted a pleasingly knobbly, branched form and chunks of these branches break off to regenerate the diploid tetrasporophyte.

OPPOSITE | The crustose species *Phymatolithon lenormandii*, named after French phycologist Sébastien Lenormand (1796–1871).

RIGHT | The maerl of northeast France, showing the stonelike appearance of many of the non-geniculate members of the Corallinales order.

JANIA

The genus *Jania*, most often represented by its species *Jania rubens*, is an example of the second geniculate kind of calcifying red seaweeds. These have gaps in their chalk coats that act as hinges, rather like the joints in plate armor. Geniculate forms seem to have evolved several times from non-geniculate ancestors, so some geniculate species are more closely related to non-geniculate species than they are to each other. This independent evolution of articulation in the chalk coating is reflected in ultrastructural differences in the way that the calcite is deposited around the non-geniculate joints of the thallus. Nonetheless, getting their reproductive cells out past their chalk coating is

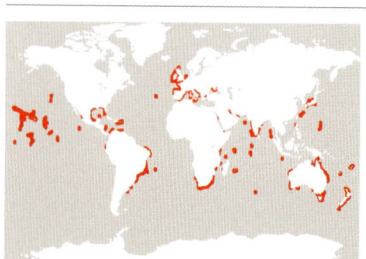

ETYMOLOGY
Named after the double-headed Roman god Ianus (Janus), perhaps because its dichotomous branching resembled Janus' two heads

DIVERSITY
Around 60 accepted species

SHAPE & SIZE
Crustose base, a few inches wide

HABITAT
Epiphytic in warmer waters worldwide; common in the eastern Mediterranean

LIFE CYCLE
Triphasic, with isomorphic gametophyte and tetrasporophyte (see page 35)

still a problem for the geniculate species and they also develop conceptacles that open into pores that the reproductive cells can pass through. In *Jania*, these conceptacles form at the tips of its fronds.

Despite their chalk coating, geniculate seaweeds are understandably more mobile than their non-geniculate relatives and look much more like we expect seaweeds to look. *Jania*, for example, grows as small, pink annual tufts with delicate white tips. It has no major commercial value but is one of the most common intertidal species in the Mediterranean Sea and has been used in traditional medicine in many Mediterranean countries.

OPPOSITE | The pink florets of *Jania rubens* (Latin: "red"), growing as epiphytes on various brown seaweeds.

ABOVE | A close-up of *Jania squamata* (Latin: "scaly"). The joints and hinges in its chalk armor are clearly visible.

AHNFELTIA

pecies of the *Ahnfeltia* genus are interesting for a number of reasons. One of them is related to the fact that Earth has, for the past few million years, followed a cycle of cold glacial and warmer interglacial periods. *Ahnfeltia* are colder-water species whose populations spread widely during the cold glacial periods and are then separated by warming waters during the interglacial periods. The geographically separated populations then develop into new species, so *Ahnfeltia* is one of the many species that can be studied to tell us something fundamental about the ways in which species evolve during ice ages.

A more mercenary reason to be interested in *Ahnfeltia* is that it is an agarophyte that forms the basis of the Russian agar industry, being found on both the Atlantic and Pacific sides of the Russian coastline, from the White Sea across to Sakhalin Island. The best known *Ahnfeltia* species, *A. plicata*, forms small, stiff, irregularly branching perennial clumps that stand erect, even when the thallus is out of the water. A couple of

ABOVE | "Wireweed." A young gametophyte of the northern Atlantic species *Ahnfeltia plicata* (Latin: "folded together").

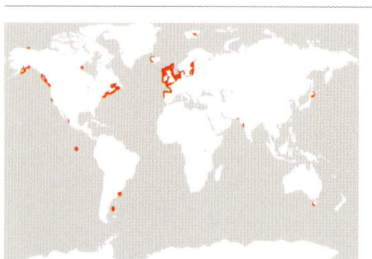

ETYMOLOGY
After Nils Ahnfelt (1801–37), a friend of Elias Fries (1794–1878) who described the genus

DIVERSITY
Around ten currently accepted species

SHAPE & SIZE
Up to 8 in (20 cm)

HABITAT
Cold temperate waters; intertidal to subtidal zone

LIFE CYCLE
Triphasic, with heteromorphic dioecious, stiff gametophyte, and crustose tetrasporophyte (see page 35)

hundred thousand tons of *A. plicata* are cultivated and harvested annually. Its use as an agarophyte supports this particular industry, but *Ahnfeltia* species have a number of more minor uses: many are eaten for either their perceived or actual health benefits, and *A. concinna*, for example, is sometimes used as a component of the Japanese thickener, funori.

A. *plicata* is also the most studied *Ahnfeltia*. The distinctive stiff thallus is the haploid gametophyte generation, which grows from a discoid holdfast and becomes fertile over the winter months. *A. plicata* shows heteromorphic alternation of generations in which the diploid tetrasporophyte is crustose and can grow to about 12 in (30 cm) across. As with

OPPOSITE, BELOW | A gametophyte of the northern Pacific species *Ahnfeltia fastigiata* (Latin: "clustered branches").

ABOVE | The stiff, upright, red gametophytes of *Ahnfeltia plicata*, scattered about with the green seaweed *Ulva*.

several other red species of this kind, the crustose diploid tetrasporophyte was originally classified as a separate species, in this case one named *Porphyrodiscus simulans*.

ASPARAGOPSIS

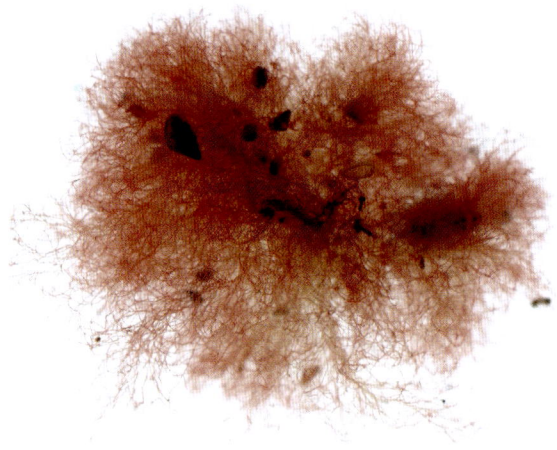

LEFT | The candyfloss-like "Falkenbergia" tetrasporophyte of *Asparagopsis armata*.

The appropriately named red *limu* of Hawaii, the haploid gametophyte generation of *Asparagopsis*, is a common, attractive, and delicate species that is often eaten in the countries near which it grows. This gametophyte generation is usually a soft pink color and its uniaxial fronds can branch either irregularly or in various paired spiral patterns. Each branched axis looks like a small, pink Christmas tree and, when seen underwater, stands of *Asparagopsis* look like miniature red pine forests. Some branches bear spines that look like barbed spears and give rise to their common name of "harpoon weed." These barbed branches do, in fact, act like fishing spears because they latch on to other seaweeds and allow *Asparagopsis* to grow alongside them.

Asparagopsis' use as a food has historically been of minor interest, but recent research may change that, because *Asparagopsis* produces a halogenated compound called bromoform ($CHBr_3$). Bromoform is probably used as a defence compound by *Asparagopsis* itself but has the interesting side effect of inhibiting bacterial methane production. This is potentially of real agricultural significance because Australian studies have shown that cattle fed on *A. taxiformis* produce much lower methane emissions, presumably because of the bromoform in those seaweed diets.

Asparagopsis shows the usual triphasic Florideophyte lifcycle. After male and female gametes fuse at fertilization, a small but visible carposporophyte generation forms on the

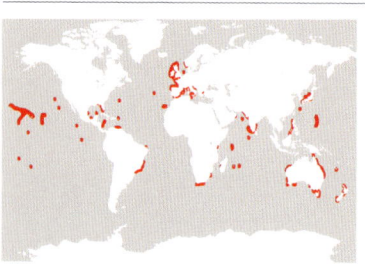

ETYMOLOGY
Greek *asparagos* = "asparagus" + *opsis* = "looking like"

DIVERSITY
Only three species are currently known

SHAPE & SIZE
Reaches around 8–12 in (10–30 cm)

HABITAT
The gametophyte is found in summer in the lower intertidal to shallow subtidal zone; the tetrasporophyte is often tangled with other species

LIFE CYCLE
Triphasic, with heteromorphic, dioecious, asparagus-like gametophyte and "Falkenbergia" tetrasporophyte (see page 35)

gametophyte. The carposporophyte is protected by a good-sized, bottle-shaped cystocarp and releases the carpospores that grow into the tetrasporophyte generation. This is very different to the asparagus-like gametophyte and looks like a small ball of candyfloss, around ½ in (1 cm) across. The tetrasporophyte is a tangled, filamentous polysiphonous mass that was originally believed to be its own species and was given the name *Falkenbergia ruflanosa*. The tetrasporophyte is sometimes referred to as the Falkenbergia stage for this reason.

RIGHT | A *Chondrus crispus* tetrasporophyte. The blade is spotted with tetrasporangia, which release the tetraspores that will regenerate the haploid gametophyte generation.

Known as "Irish moss," *Chondrus* species are rich in vitamins and eaten worldwide: *C. elatus* is common in southeast China and *C. ocellatus* in Japan, but the best-known species in Europe and Canada is *C. crispus*, harvested for carrageenan in Europe, and whose extracts have been used as cooking thickeners in Ireland for hundreds of years.In fact, the word "carrageen" comes from the Irish name for *C. crispus*, which is *carraigin*.

Carraigin translates into English as "little rock" and it is not a bad description of what *Chondrus* species look like along many coastlines. They are relatively small, with a discoid holdfast and a short stipe that opens into a flat blade with regular dichotomous branches. This can look like a small, red, symmetrical rock. They grow in a range of habitats, can support epiphytes, and can grow as epiphytes themselves on kelps. Growth is environmentally sensitive, so species can show a wide range of growth forms (thinner or thicker blades, more or fewer divisions), which makes them hard to identify. Thalli grow mainly in spring and summer:

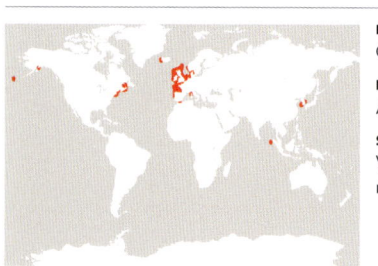

ETYMOLOGY
Greek *chondros* = "cartilage"

DIVERSITY
Around 40 accepted species

SHAPE & SIZE
Very variable, but dichotomous and usually not much larger than hand sized

HABITAT
Low intertidal and subtidal zone

LIFE CYCLE
Triphasic, with isomorphic gametophyte and tetrasporophyte (see page 35)

the diploid carposporophyte grows in summer and the diploid tetrasporophyte releases spores in spring and summer. The haploid gametophytes are dioecious and there is some minor sexual dimorphism: the haploid male gametophytes have white patches at the tips of their fronds while the haploid female gametophytes have bumplike procarps that develop into visible carpospores after fertilization.

Chondrus shows isomorphic alternation of generations. The haploid gametophyte and diploid tetrasporophyte are the same shape but show biochemical differences: the haploid gametophyte cell wall contains more κ-carrageenan than the diploid tetrasporophyte cell wall, for example, although exactly why remains unclear. The

C. crispus genome was sequenced in 2013 to help address questions of this sort and, as is always the way, raised more questions that we are still exploring.

BOTTOM | "Irish moss," *Chondrus crispus* (Latin: "curled"), growing on a pebbly shore in a broad, flat shape.

BELOW | Dried Irish moss, also known as carrageen moss. This handful will be used to make herbal medicine.

GIGARTINA

BELOW | The slender branching gametophyte generation of *Gigartina pistillata* (Latin: "club-shaped").

S pecies of the *Gigartina* genus display a broad range of shapes and sizes, so the genus has historically seen a lot of taxonomic traffic. The type specimen, Linnaeus' 1759 *Fucus gigartinus*, was renamed *F. pistillatus* later in the eighteenth century

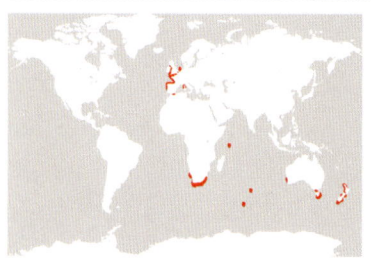

ETYMOLOGY
Greek *gigarton* = "grape stone"

DIVERSITY
Around 80 species

SHAPE & SIZE
Very variable, but dichotomous and often hand sized. Some species can be much larger

HABITAT
Intertidal and subtidal zone

LIFE CYCLE
Triphasic, with heteromorphic gametophyte and tetrasporophyte (see page 35)

and then given its own genus by John Stackhouse in 1809 to become *G. pistillata*. The genus currently contains around 50–100 species but was, for a time, one of the default genera into which to place carrageenophytes, so around twice that number of species have been proposed and eventually moved into other genera. The common *Mastocarpus stellatus*, for example, was for many years classified as *Gigartina stellata*. Similarly, the commercially important South American carrageenophyte *Sarcopeltis skottsbergii* (see page 122) was, until recently, classified as *Gigartina skottsbergii*.

The representative *G. pistillata* reaches perhaps 8 in (20 cm) in length, with heteromorphic generations. It is edible and a number of credible claims have been made for the health benefits of its various extracts. The visible generation is the haploid gametophyte, which forms a straplike, fleshy, dichotomously branching frond that tapers rather delicately at its tips. A long stipe connects it to a discoid holdfast and it is relatively common on the shorelines of Atlantic Europe. The diploid tetrasporophyte is much smaller and grows hidden inside the gametophyte. However, in many areas, *Gigartina* populations appear to reproduce asexually through the female gametophyte rather than sexually. As with most reds, colors can vary from red to almost black, depending on conditions, but a rich crimson color is not uncommon and can be very beautiful.

SARCOPELTIS

The *Sarcopeltis* genus is currently known only from the coastlines of southern South America and the Antarctic Peninsula and contains two species: *S. skottsbergii*, which is the most important carrageenophyte in the Chilean carrageenan industry, and *S. antarctica*. However, *Sarcopeltis* is one of the most recently described genera in this book, so more species may be out there to be added: the current two were both called *Gigartina skottsbergii* for more than 80 years, with their move to an entirely new genus, *Sarcopeltis*, coming in 2020 after DNA-based reconsideration of their phylogenetic positions.

S. skottsbergii is one of the three *lugas* of Chile, together with *Sarcothalia crispata* and *Mazzaella laminaroides*, both of which are in the same family as *S. skottsbergii*, the Gigartinaceae. All three species grow in the sub-canopy under the giant kelp, *Macrocystis pyrifera* (see page 215), and are harvested together. Recent years have seen problems with overextraction and the industry in Chile is increasingly looking at cultivation and management to maintain sustainable crop yields.

The thalli are some of the largest seen in the reds, which is one of the reasons why they make good carrageenophytes. *Sarcopeltis* species show isomorphic alternation of generations, with the gametophytes and tetrasporophytes being leathery and circular and sometimes reaching 40 in (100 cm) or so across. One characteristic feature of the new *Sarcopeltis* genus is that these thalli can produce additional, secondary holdfasts, which may be an adaptation to cope with the large body of the seaweed. Another characteristic feature is that tetrasporangia in this genus are formed from filaments in the medulla, rather than in the cortex. Both haploid and diploid generations may be found growing together at the same time, although there is some evidence that populations contain more haploid gametophytes in winter and more diploid tetrasporophytes in summer.

OPPOSITE | The heavy thallus of *Sarcopeltis skottsbergii*, named after the Swedish phycologist Carl Skottsberg (1880–1963). This thallus is dimpled with reproductive structures (probably carposporophytes) and is growing in the Falkland Islands.

ETYMOLOGY
Greek *sarko* = "flesh" + *pelte* = "shield"

DIVERSITY
Two currently known, but this is a new genus

SHAPE & SIZE
Large, leathery, oval fronds that can be 40 in (100 cm) across

HABITAT
Deeper subtidal waters off western South America

LIFE CYCLE
Triphasic, with isomorphic gametophyte and tetrasporophyte (see page 35)

GLOIOPELTIS

LEFT | *Gloiopeltis furcata* (Latin: "fork-shaped"), with the green seaweed *Ulva* in the background.

ETYMOLOGY
Greek *gloio* = "sticky" + *pelte* = "shaft"

DIVERSITY
Around half a dozen known species

SHAPE & SIZE
Tubular branches, usually the size of a small shrub

HABITAT
Sandier shores

LIFE CYCLE
Triphasic, with isomorphic gametophyte and tetrasporophyte (see page 35)

Known as "glueweed," *Gloiopeltis* is the major genus used in the production of funori, with the major species being *G. furcata*. Funori is an agar-like paste that has been used to either glue or stiffen fabrics in Japan since at least the seventeenth century. Its most familiar application is probably as one of the adhesives on Japanese *shōji* (paper screens and doors), but it has been widely used: as a glaze for porcelain, as a cleaning agent for silk kimonos, as a bookbinding glue, and, more recently, as a conserving agent for delicate museum artifacts.

Gloiopeltis shows isomorphic alternation of haploid gametophyte and diploid tetrasporophyte generations, with the diploid carposporophyte being restricted to cystocarps on the female gametophyte. Gametophytes and tetrasporophytes are small and multiaxial, with flat blades and smooth, cylindrical branches that grow from a small discoid holdfast. The genus grows year-round, mainly in the warmer waters of the North Pacific, and is eaten in coastal communities in China.

Gloiopeltis may be harvested from wild populations but coastal communities in both China and Japan have also cultivated it for the better part of a millennium. Rockeries are made just above the low-water mark to give the *Gloiopeltis* something to hold onto and these are then seeded in the spring. *Gloiopeltis* grows fast, so it may be harvested later in the year.

Historically, funori manufacture centered around the southern Japanese port of Osaka. After the fronds were harvested, they were cleaned and then soaked or steamed in freshwater to soften them, before being stretched into thin layers that were dried slowly to keep them flexible. Once ready, these were spooled into the large rolls that were traditionally sold. When needed, the funori was then easy to make simply by peeling off a length from the roll and dissolving it in hot water to give the paste.

OPPOSITE | A *Gloiopeltis furcata* rockery on Awaji Island, on the south coast of Hyogo prefecture, Japan.

ABOVE | A closer view of *Gloiopeltis furcata* fronds growing on the rockery on Awaji Island.

FURCELLARIA

This seaweed is a rare example of a monotypic genus, which is to say that it only contains one species: *Furcellaria lumbricalis*. This species does, however, have two growth forms: one is attached and one is free-floating. The free-floating form was originally believed to be a separate species and was first described by Linnaeus in 1753 as *Fucus fastigiata*. It was then renamed *Furcellaria fastigiata* before being brought into *F. lumbricalis*.

Furcellaria grows across a wide region of the North Atlantic but is perhaps best known as the dominant red seaweed in the Baltic Sea. It is perennial and impressively tolerant of the Baltic's low salinities. The free-floating form commonly forms large offshore mats, which reproduce by vegetative propagation. In Denmark, and more recently in Canada, these mats are collected and used to produce a substance called furcellaran, which is a type of carrageenan. The attached growth form grows from very branched rhizoidal holdfasts, which often develop into long and elegant stipes before branching and fronds can grow to around 20 in (50 cm) long. The fronds themselves are long and fleshy and will dichotomously branch several times at their ends, earning the species the common name of "clawed fork weed."

In contrast to the vegetative propagation of the free-floating form, the attached form of *F. lumbricalis* reproduces sexually and shows almost isomorphic alternation of generations: the diploid tetrasporophytes look the same as the haploid gametophytes but are a little larger. When fertile, cylindrical reproductive structures form at the ends

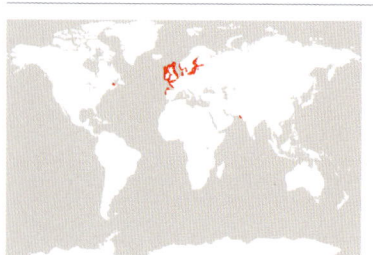

ETYMOLOGY
Latin *furcilla* = "little fork"

DIVERSITY
Only one known species

SHAPE & SIZE
Forked and bushy

HABITAT
North Atlantic and its marginal seas; grows on rockier shores or in floating mats

LIFE CYCLE
Triphasic, with isomorphic gametophyte and tetrasporophyte (see page 35)

of the clawed branches. The female gametangia and the diploid tetrasporangia look similar, while the haploid male spermatangia are more distinctive. Populations are a mix of haploid gametophytes and diploid tetrasporophytes, although the diploid tetrasporophytes are usually more common than the haploid gametophytes.

OPPOSITE | Forkweed, or *Furcellaria lumbricalis* (Latin: "wormlike").

The *Hypnea* genus includes important crop carrageenophytes that are harvested for κ-carrageen as well as a number of species that are eaten as foods. *H. nidifica* is the "hidden *limu*" of Hawaii, for example, which is used as both food and medicine. *Hypnea* species are also eaten in Indonesia and used as fertilizer in some countries in South America. They occur in significant volumes, and are increasingly farmed or harvested, in coastlines along the northern Indian Ocean, across to the Persian Gulf and down the coast of east Africa. Overall yields of carrageenan remain small compared to *Eucheuma* and *Kappaphycus*, but *Hypnea* holds promise as a crop in areas where the other two genera do not grow.

Hypnea species show isomorphic alternation of generations: the haploid gametophyte and diploid tetrasporophyte consist of irregular cylindrical branches that grow into tangled masses that can end up covering decent amounts of any beach. However, while species can reproduce sexually, they can also reproduce asexually, and population structures remain unclear in some places.

Some species can also grow by vegetative propagation: fragments of a thallus can break off and grow into new thalli. A potentially problematic species is *H. musciformis*, or "hookweed," which is known to propagate in this way. As its common name suggests, *H. musciformis* has small hooks at the ends of its filaments, which allow it to latch on to other species and grow as an epiphyte. This species was introduced into Hawaii as a crop carrageenophyte in the mid-twentieth century and has rapidly escaped and expanded to become an invasive that is outgrowing the native *H. cervicornis*. The effects of this invasion may not be entirely bad:

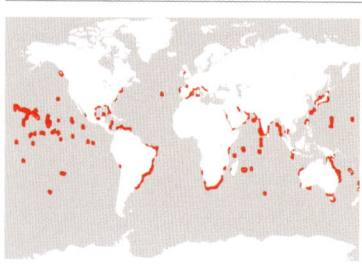

ETYMOLOGY
Greek *hypnos* = "sleep"

DIVERSITY
Around 60 accepted species

SHAPE & SIZE
Bushy, tangled mats that can be 40 in (100 cm) or so in length

HABITAT
Intertidal and shallow subtidal zone

LIFE CYCLE
Triphasic, with isomorphic gametophyte and tetrasporophyte (see page 35)

H. musciformis grows so fast that it can create large floating offshore mats, which have become part of the diet of Pacific green sea turtles.

OPPOSITE | Tangled fronds of *Hypnea spicifera* (Latin: "tufted").

RIGHT | The hooks of hookweed, *Hypnea musciformis* (Latin: "mosslike").

EUCHEUMA

The *Eucheuma* genus, along with *Kappaphycus* (see page 130), is one of the two main genera used for the production of carrageenan. Species in these genera show very variable growth and can therefore look very similar, but they only share their order and their family, not their genus. *Eucheuma* itself is a tropical genus found most commonly around the shorelines of the western Pacific and Indian oceans. The main *Eucheuma* species is now called *E. denticulatum*, although historically it was known as *E. spinosum* and is still often referred to by that name. It is an excellent source of κ-carrageenan and an excellent crop: it grows fast and reliably from an apical meristem, often reaching 40 in (100 cm) or so in length after only a couple of months' growth. It also grows well in shallow, sandy, easy-to-wade-out-to areas of the upper subtidal zone, which makes it relatively easy to harvest. The vast majority is grown in Indonesia and the Philippines, but its farms are cheap to set up and substantial *E. denticulatum* farms have been established as far north as China, as far west as Tanzania on the east coast of Africa, and as far east as Kiribati in Oceania.

The carrageenan industry is far and away the largest market for the perhaps 5 million tons of *Eucheuma* harvested worldwide every year, but species are also eaten. *Eucheuma* species are, for example, used in *gusô*, a kind of relish, in traditional Philippine cuisine. *Eucheuma* usually grows as fleshy, tendril-like thalli with irregular branches and small spines running along the edges of the thalli.

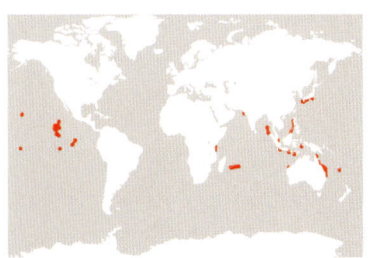

ETYMOLOGY
Greek *eu* = "true" + *cheuma* = "basin"

DIVERSITY
About 20 species are known

SHAPE & SIZE
Bushy, with larger individuals reaching around 40 in (100 cm) in length

HABITAT
Shallower subtidal zone

LIFE CYCLE
Triphasic, with isomorphic (dioecious) gametophyte and tetrasporophyte (see page 35)

However, they show a lot of growth variation and the crop is important enough that breeders are beginning to cultivate new, higher-yield, strains.

Eucheuma life cycles are reliably triphasic, with isomorphic haploid gametophyte and diploid tetrasporophyte generations. Following fertilization, the genus has a slightly unusual method of carposporophyte development in which more cells than usual in the fertilized carpogonium fuse together to form a large fusion cell. This fusion cell then develops into the diploid carposporophyte. The surrounding cells form a dense protective pericarp and the distinctive carposporophyte may be seen projecting from the thallus, often on small stalks.

OPPOSITE | The branched fronds of *Eucheuma denticulatum* (Latin: "toothed").

ABOVE | A *Eucheuma* farm in Zanzibar, Tanzania. Cuttings are tied to ropes, staked out, and left to grow in shallow bays that make for easier harvesting.

KAPPAPHYCUS

This is the second of the two genera cultivated as the major source of carrageenan in Indonesia and other countries around the Indian Ocean and Western Pacific (see also *Eucheuma*, page 130). The most common species, *Kappaphycus alvarezii*, is the old *Eucheuma cottonii* and is an excellent source of κ-carrageenan. *K. alvarezii* makes one of the largest contributions of any single species to global carrageenan production.

Kappaphycus species are some of the largest of the red seaweeds, reaching 6 ½ ft (2 m) in length after a couple of months. The fronds are fleshy and multiaxial with irregular branching, to give a bushy appearance, and thick, tangled mats are common. While *Kappaphycus* is a "red" seaweed, it tends to be more orange or green in color. Growth is very plastic, however, and many fronds are smaller and more delicate, depending on growth conditions.

LEFT | *Kappaphycus alvarezii*, named by the US biologist Maxwell Doty (1916–1996) for his colleague, Vicente Alvarez. Differently colored cultivated strains of *Kappaphycus* are shown here in Florianópolis in Santa Catarina, Brazil.

OPPOSITE | A *Kappaphycus* harvest drying on Kangge, one of southeast Indonesia's lesser Sunda Islands.

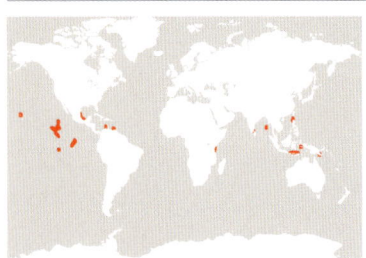

ETYMOLOGY
The letter kappa represents the type of carrageenan it supplies (κ-carrageenan) + Greek *phycos* = "seaweed"

DIVERSITY
About half a dozen known species

SHAPE & SIZE
Bushy and can reach 6½ ft (2 m) in length

HABITAT
Tropical, often found around coral reefs in the low intertidal and subtidal zones

LIFE CYCLE
Triphasic, with isomorphic gametophyte and tetrasporophyte (see page 35)

The genus shows isomorphic alternation of generations and natural populations can be a mix of the haploid gametophytes and the diploid tetrasporophytes. As with *Eucheuma*, *Kappaphycus* grows well in the sheltered and shallow subtidal zone, making it an extremely good crop and it is indeed widely cultivated. Notably, *Kappaphycus* grows very well after vegetative propagation. This is very useful for farming, and plantations are usually started by tying cuttings to lines along the subtidal zone. The fronds may then be easily harvested at low tides a few months later.

The ease and speed of *Kappaphycus* cultivation has led to a number of initiatives that aim to provide alternative economies to coastal populations around the Indian Ocean and, especially, to provide employment to women in areas that have more traditionally relied on male fishermen: farms in Tamil Nadu and Tanzania bear some witness to this.

The speed with which *K. alvarezii* can grow by vegetative propagation does, however, make it a pernicious invasive species and large, wild, invasive populations have become established in Hawaii after the species was introduced for cultivation in the 1970s. These invasives can overshadow indigenous corals, although they can also provide new habitats that support local fish, so the extent to which, on balance, invasive *Kappaphycus* might damage ecosystems worldwide is still unclear.

BELOW | *Kappaphycus* lines growing offshore in Kangge, one of southeast Indonesia's lesser Sunda Islands. The shallow waters make harvesting the seaweed easier.

GELIDIUM

Species in the *Gelidium* genus are eaten worldwide, are relatively common, are a good source of agar, and are often rather beautiful, so are a popular species for amateur naturalists to collect and preserve. Perhaps relatedly, the genus contains one of the largest species counts of any of the red seaweed genera.

Gelidium shows isomorphic alternation of generations. The population structures are still being investigated, but cooler waters may favor the haploid gametophytes while warmer waters may favor the diploid tetrasporophytes. There is considerable morphological variation within the genus, but species are generally perennial and grow on rocks or as epiphytes from a horizontal filament. This gives rise to an erect branched axis that develops into flattened pinnate fronds that have smaller branchlets. The reproductive structures are found on these branchlets and some species reach around 20 in (50 cm) in length.

RIGHT | The delicate fronds of *Gelidium spinosum* (Latin: "spiny").

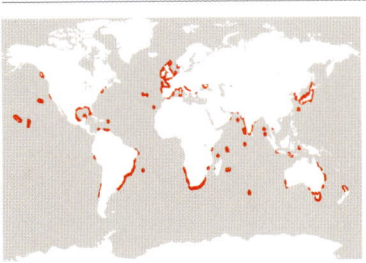

ETYMOLOGY
Greek *gelidus* = "congealed" (because they were once boiled into gels)

DIVERSITY
Around 150 currently accepted species

SHAPE & SIZE
Very variable, but often bushy and around 20 in (50 cm) in length

HABITAT
Rocky shores

LIFE CYCLE
Triphasic, with isomorphic gametophyte and tetrasporophyte; species may be mono- or dioecious (see page 35)

The commercial value of the genus is based largely around one species, the agarophyte *G. amansii*. Wild harvested *Gelidium* was historically an important source of agar, although cultivated *Gracilaria* has since taken over. Agar is a Malay word, but the industry in its modern form began in Japan around the middle of the seventeenth century and was centered around Hokkaido in the north and Osaka in the south. Most *Gelidium* is still harvested from wild populations, rather than being farmed, and today's demand for agar means that many other countries, most notably Morocco, now collect significant amounts of *Gelidium* biomass.

The foundations of the agar industry go back to at least the early medieval period, and to a kind of noodle that was probably first made in China and brought to Japan. Farmers and fishermen would

BELOW | The fine red fronds of *Gelidium attenuatum* (Latin: "slender") growing next to the thicker fingers of the green seaweed *Codium tomentosum* on the shoreline of Galicia in northwest Spain.

rake up *Gelidium* species or dive to collect them. The *Gelidium* would be washed, cleaned, acidified, boiled, strained, and finally congealed into noodles that, in Japan, were called *tokoroten*.

Tokoroten is delicious, but perishable. Around 1650, a perhaps apocryphal tale introduces a Kyoto innkeeper called Tarozaemon Minoya. Various versions of the story exist but, for some reason or for some dignitary, the innkeeper prepared too much *tokoroten*. Realizing his mistake, he threw the surplus into the snow only to find, the next morning, that it had frozen and dried into a white, nonperishable block to leave what is now called *kanten*. The word *kanten* is etymologically ambiguous and is likely to be a contraction of the Japanese words for "cold seaweed," but the substance was what we now call agar and this freeze-drying method is still in use today.

BELOW | Gathering *Gelidium sesquipedale* (Latin: "a foot and a half") for agar extraction, off the coast of Asturias in northern Spain.

GRACILARIA

A warm to temperate water genus, *Gracilaria* contains several species that are either eaten, especially as abalone feed, or have been used as agarophytes. *G. gracilis* is a particularly important agar species.

The genus is another large one and its species display a comparably wide range of morphologies: larger *Gracilaria* species can reach perhaps 20 in (50 cm) in length. Fronds are usually uniaxial (that is, they have one main axis), which grows from a discoid holdfast. The main axis branches early and often, so it quickly becomes all but impossible to distinguish it from its branches. The branching is sub-dichotomous or irregular and the fronds may be long and slender (for example, *G. gracilis*) or broader and flatter (for example, *G. multipartita*).

At the ultrastructural level, the carpogonia on the female gametophytes are two-celled, with a supporting cell. After fertilization, these develop into rounded cystocarps that are clearly visible and have pores that are much harder to see and through which the carpospores are released. These wart-like cystocarps give members of the genus their understandable common name of "wart weeds."

LEFT | The warts of wartweed, in this case *Gracilaria multipartita* (Latin: "many divisions"). The warts are the cystocarps, which will release the carpospores that grow into the tetrasporophyte.

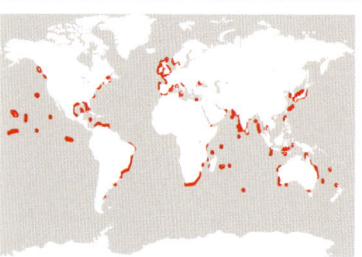

ETYMOLOGY
Latin *gracilis* = "slender"

DIVERSITY
Around 150–200 currently accepted species

SHAPE & SIZE
Bushy, reaching maybe 20 in (50 cm) in length

HABITAT
Grows on rocks in the intertidal and subtidal zones, especially in shallow bays. It can tolerate a wide range of conditions, however, so grows opportunistically

LIFE CYCLE
Triphasic, with isomorphic gametophyte and tetrasporophyte; isomorphic dioecious (see page 35)

LEFT | *Graciliaria gigartinoides* (Greek: "grape stonelike") growing on the west coast of Vancouver island, British Columbia, Canada.

During the development of the worldwide agar industry, many countries realized that they did not have many *Gelidium* species but did have *Gracilaria* species. They also realized that *Gracilaria* could be used to produce agar and that *Gracilaria* was easier to farm than *Gelidium*, although its farms are perhaps more susceptible to an oömycete pathogen that causes the appropriately named red rot. *Gracilaria* species have therefore taken over from *Gelidium* as the most important species involved in agar production. Around 4 million tons of cultivated *Gracilaria* are harvested each year, almost all of it in China. This works out to around 10 percent of global cultivated seaweed yields, making *Gracilaria* an extremely important crop genus.

GRATELOUPIA

LEFT | The flat blades of *Grateloupia capensis* (Latin: "found at the cape").

Various *Grateloupia* species have been used and eaten worldwide for hundreds of years. The genus is diverse and its species show a lot of variety. Fronds grow from a discoid holdfast and may be flat or cylindrical, often growing to lengths of around 20 in (50 cm). The genus shows isomorphic alternation of generations and there is evidence that different generations are favored at different times of the year. The various species around Japan are used as a minor component of funori (see also, page 125) and are eaten around the Pacific, including as *limu* in Hawaii and *kata-nori* in Japan.

Unfortunately, the good reputation that the genus *Grateloupia* has slowly built for itself has been spoiled recently by its worst behaved member, the highly invasive "devil's tongue weed," *G. turuturu*.

G. turuturu is a perennial that is native to the northwest Pacific, but it also has all the traits to make

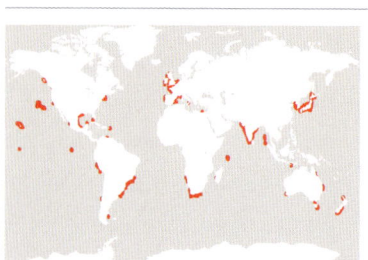

ETYMOLOGY
Named after Jean-Pierre de Grateloup (1782–1862), a French naturalist who knew many of the nineteenth-century French phycologists

DIVERSITY
About 100 currently accepted species

SHAPE & SIZE
Variable, but some species can reach 6½ ft (2 m) in length

HABITAT
Worldwide subtidal zone; many species prefer sheltered areas, and some (e.g. *G. turuturu*) are opportunistic

LIFE CYCLE
Triphasic, with isomorphic (monoecious) gametophyte and tetrasporophyte (see page 35)

it an excellent invasive species: it grows quickly in nutrient-rich water and survives well across a range of temperatures and salinities. *G. turuturu* thalli are multiaxial, with each axis growing into a broad blade that can be over 40 in (100 cm) in length. Its thalli are both fertile and able to spread well through vegetative propagation and *G. turuturu* appears to have done exactly that over the past decade or two. Its populations show considerable genetic homogeneity, which suggests that they came from a single area by a single route. Our best current idea is that *G. turuturu* spread as a hitchhiker in the shellfish shipments that support mollusk aquaculture. However it happened, *G. turuturu* is now found worldwide and seems to be acting as an opportunistic destroyer of habitats. When other red species decline, perhaps driven by warming waters, *G. turuturu* becomes established and grows rapidly. Its large size allows it to overshadow other, smaller reds, eventually outcompeting them.

BELOW | The invasive *Grateloupia turuturu* (from the Japanese name for this species, which is *turuturu*) seen here spreading over the native brown seaweed *Bifurcaria bifurcata* off the coast of Galicia, Spain.

BOTRYOCLADIA

Surprisingly, *Botryocladia botyroides* was not included in Linnaeus' 1753 *Species Plantarum*. It did, however, find its way into his 1759 *Systema Naturae*. The surprise is because the genus is an extremely distinctive one, whose species are well served by their common name of "sea grapes."

Species are found worldwide, although usually in warmer waters. *Botryocladia* is subtidal and species often form part of the sub-canopy in kelp forests. The thalli are also relatively small, not growing much more than 8–12 in (20–30 cm) in length, which makes them reasonably common aquarium species.

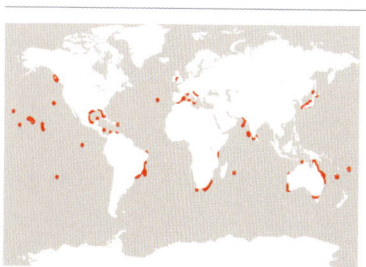

ETYMOLOGY
Greek *botrys* = "grapes" + *klados* = "branch"

DIVERSITY
Around 50 known species

SHAPE & SIZE
Distinctive, grape-like appearance, usually small and about hand sized

HABITAT
Warmer subtidal waters worldwide

LIFE CYCLE
Triphasic, with isomorphic gametophyte and tetrasporophyte; dioecious (see page 35)

Life cycles in the *Botyrocladia* genus are isomorphic. The larger gametophyte and tetrasporophyte phases consist of branching axes. The smaller branches of the thallus have tubular lower sections that expand into large, mucilage-filled sacs: the "grapes." These sacs vary in size between species. As an example, the colder-water species *B. pseudodichotoma*, which is found along the American and Canadian Pacific, has branches that end in sacs that are an inch or so across. The warmer-water *B. occidentalis* of the western Atlantic has sacs less than half that size. The full function of *Botyrocladia* sacs remains a matter of some discussion, but they can certainly act as flotation bladders and may play other roles in metabolism.

OPPOSITE | "Sea grapes," *Botryocladia sonderi*, named after the German botanist Otto Sonder (1812–1881) and seen here growing off the coast near Perth, Western Australia.

TOP | The red sea grapes of *Botryocladia leptopoda* (Greek: "thread foot") also growing off Perth, Western Australia.

ABOVE | The large sacs of *Botryocladia pseudodichotoma* (Greek: "falsely forking") can be up to 1 in (2 cm) across.

CERATODICTYON

The genus *Ceratodictyon* is of interest because of the weirdness of one of its species: the "holey sponge" seaweed, *C. spongiosum*. At first glance, *C. spongiosum* looks roughly seaweed-shaped: the fronds are broad and flattened and apparently dichotomously branched. As their common name suggests, they look spongy. The thalli are found in the high subtidal zone of warmer waters near coral reefs and are often more green than red.

However, at second glance these fronds turn out to be not one species, but two, and the second species is not a seaweed, but the sponge *Haliclona cymaeformis*. The whole frond, therefore, is a symbiosis between a red seaweed and a sponge, which is, unsurprisingly, why it looks spongy and why it is found near coral reefs.

In fact, the seaweed partner, the actual *C. spongiosum*, grows as a meshlike network of filaments, over and between which the sponge partner grows in a manner that is reminiscent of the association between algae and fungi in a lichen. The whole symbiosis usually reproduces by vegetative propagation from its fragments, but reproductive structures have been observed on the algal partner and may be used to classify it (for example, the presence and nature of the cystocarp and fusion cells). The *C. spongiosum* partner can live independently of the sponge in culture but has not been seen to do so in the wild. The sponge partner, on the other hand, dies if separated from the alga.

The exact purpose of this symbiosis remains a matter of some speculation, although both partners are known to work at it because they both produce chemical signals that help to maintain the symbiosis. The seaweed appears to use the nitrogen contained in the waste products of the sponge, but the sponge continues to filter feed, so the benefits to the sponge may be more structural than nutritional: the seaweed may simply help to anchor the sponge in place.

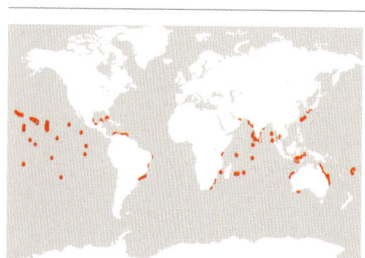

ETYMOLOGY
Greek *kerastes* = "horned" + *dictyon* = "a net"

DIVERSITY
About ten species

SHAPE & SIZE
Variable, but often branched and shrub sized

HABITAT
Often found near coral reefs

LIFE CYCLE
Triphasic (see page 35)

ABOVE | Holey sponge seaweed, the extraordinary *Ceratodictyon spongiosum* (Latin: "spongy"), seen here on the shore of Lizard Island in northeast Queensland, Australia.

PLOCAMIUM

Species of *Plocamium* are widely found in temperate upper subtidal waters. They are often a crimson or bright red color and commonly grow on rocks or as an epiphyte on larger seaweeds. The fronds are erect and flattened, growing from a discoid holdfast and reaching perhaps 8 in (20 cm) in length. The haploid and diploid generations are isomorphic and are often found growing together, although some evidence suggests that certain seasons favor certain generations.

Members of the *Plocamium* genus are beautiful close up, but their beauty can be harder to see at the shoreline because they are highly branched and often grow in tangled clumps. They transform under magnification: the frond branches and each branch supports a delicate and translucent comb, hence their common name, "cock's comb." The reproductive sporangia and gametangia develop on these branchlets and, after sexual reproduction between gametophytes, globular cystocarps develop on the same branchlets and are visible to the naked eye.

The *Plocamium* genus is scientifically interesting as an example of cryptic diversity. The species *P. cartilagineum* was, for a long time, believed to be one species, but "*P. cartilagineum*" individuals have recently been shown to belong to at least four

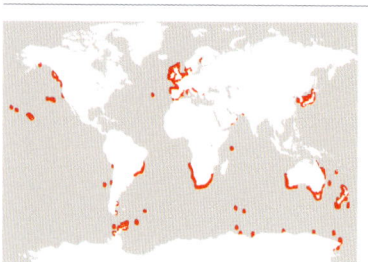

ETYMOLOGY
Greek *plokamos* = "a lock of hair"

DIVERSITY
About 50 currently accepted species

SHAPE & SIZE
Branched and tangled, reaching perhaps 8 in (20 cm) in length

HABITAT
Intertidal zone worldwide

LIFE CYCLE
Triphasic, with isomorphic gametophyte (dioecious) and tetrasporophyte (see page 35)

genetically distinct species (*P. cartilagineum*, *P. subtile*, *P. nanum*, and *P. maggsiae*). This raises a number of questions, because these four species are essentially impossible to tell apart by eye. What differences, therefore, lead to the development of new species? The short answer is that we do not know, which makes this an exciting field for investigation.

OPPOSITE | The beautiful red fronds of *Plocamium rigidum* (Latin: "stiff").

TOP | Tangled mats of *Plocamium maggsiae* growing on the south coast of Portugal; named after Christine Maggs, an extant British phycologist.

ABOVE RIGHT | *Plocamium nanum* (Latin: "dwarf"), showing the delicate combs that give species in this genus the name of "cock's comb."

CERAMIUM

The genus *Ceramium* contains some species that have minor commercial uses (for example, *C. boydenii* is a minor component of agar), but we include them here because their beauty is matched by, and draws from, some interesting biology. The genus is one of the largest in the red algae, with around 300 currently accepted species, but is also one of the taxonomically messiest, with many species needing reevaluation. Species assignations with the *Ceramium* genus are therefore likely to change in the not-too-distant future.

Most *Ceramium* species are relatively small and usually shorter than around 12 in (30 cm) in length. They have cylindrical fronds that are often very

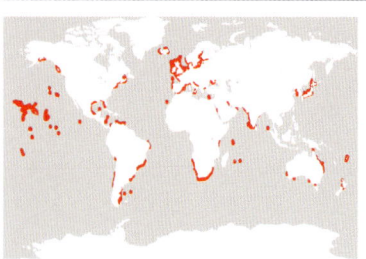

ETYMOLOGY
Greek *keramium* = "a vessel"

DIVERSITY
Around 300 accepted species

SHAPE & SIZE
Bushy and shrub sized, not usually more than 12 in (30 cm) long

HABITAT
Intertidal and subtidal zone worldwide, often as epiphytes

LIFE CYCLE
Triphasic isomorphic alternation of generations (with gametophytes usually dioecious)

branched to give bushy thalli, but their small size and large diversity can make it hard to tell one species from another. Thalli can grow on rocks or as epiphytes and the tips of their fronds usually end in distinctive, pincerlike, paired branches. One of the common names for *Ceramium*, appropriately, is "pincer weed."

The architecture of *Ceramium* fronds is interesting for two reasons. First, some *Ceramium* species can look banded. This is because a single row of large cells forms the main axis of each frond, but an outer layer, or cortex, of smaller cells then grows over these large cells. The cortical cells can either grow in a continuous layer all the way along the frond or they

OPPOSITE | An epiphytic species of *Ceramium* growing on the brown seaweed, *Himanthalia*, off the coast of Cornwall, southern UK.

ABOVE | The red fronds of a delicate *Ceramium* species growing alongside the flatter, forked fronds of the brown seaweed *Dictyota dichotoma* off the Spanish Mediterranean coast.

can grow in clusters over the nodes where two large main-axis cells meet. If they grow over the nodes, then the cortical cells appear darker than the large main-axis cells and this gives some *Ceramium* species their distinctive banding patterns: this is called cortication. *Ceramium* shows isomorphic alternation of generations and the reproductive structures in each generation form on the cortical bands, although they are too small to see by eye.

The second interesting thing about their architecture is that a species' branching often looks dichotomous, but this dichotomy is produced by a different mechanism to the dichotomy of other seaweeds. The branches of *Ceramium* do not divide symmetrically into equal-length branches (that is, they are not truly dichotomous). Instead, a lateral branch develops from the main axis but, in a process called evection, the branch and main axis then rotate away from each other. This gives a more symmetrical branch, so we say that *Ceramium* is pseudo-dichotomous (see page 79).

TOP | *Ceramium echionotum* (Greek: "snake-backed") growing as an epiphyte.

ABOVE | The pincerlike tips that give many *Ceramium* species their common name of pincer weeds. The distinctive banding, or cortication (see main text), seen in some species is clearly visible.

PTEROTHAMNION

RIGHT | The delicate fronds of *Pterothamnion plumula* (Latin: "feathery").

The genus *Pterothamnion* is another beautiful one. Fronds are rose-red to pink and, as with *Ceramium* (see page 148), they are pseudo-dichotomous, growing to around 8 in (20 cm). The main axes and primary branches are monosiphonous and the primary branches bend upward and inward. Two to four smaller branches come off each main axis or primary branch cell. Those smaller branches can be arranged around the axial cells in a variety of patterns: they may be paired opposite to each other, or may alternate, or may be at right angles to each other. The smaller branches also grow branchlets of their own and the reproductive cells form on these branchlets. The overall effect is to make the *Pterothamnion* frond look like a feather and a common name for this genus is, appropriately, the "feather weeds."

Species in this genus have a variety of growth forms and are widely distributed. Their architecture, however, is relatively poorly understood: it used to be thought that the development and patterns of the branches and branchlets could be used to distinguish one species from another. Recent DNA-based molecular phylogenies, however, suggest that this may not be the case and that we need to reconsider the number of species in this genus.

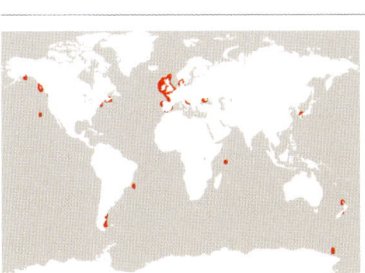

ETYMOLOGY
Greek *ptero* = "feather" + *thamnos* = "a shrub"

DIVERSITY
About 30 currently accepted species

SHAPE & SIZE
Branched and feathery, usually about hand sized

HABITAT
Intertidal and subtidal zone

LIFE CYCLE
Triphasic isomorphic alternation of generations (with gametophytes usually dioecious; see page 35)

ACANTHOPHORA

The edible *bulung tombong bideng* of Indonesia, the genus *Acanthophora* was represented by *A. muscoides* in Linnaeus' 1753 *Species Plantarum* and is perhaps best known today for its invasive member, *A spicifera*.

The genus shows a lot of growth variation both between and within species but fronds have distinctive spines that give them their common name of "spiny red seaweed." Species grow erect from a disk-shaped holdfast and the fronds are fleshy and often reasonably cylindrical. The main axis has no spines, but there is much irregular branching and the main branches do have spines. When fertile, the reproductive structures form on these spines.

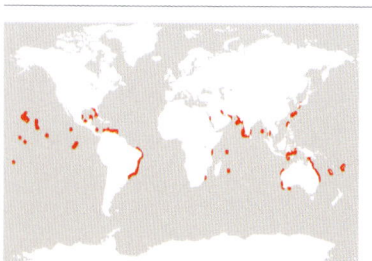

ETYMOLOGY
Greek *acantho* = "thorn" + *phoros* = "bearer"

DIVERSITY
Around five to ten species are currently accepted

SHAPE & SIZE
Small and spiky

HABITAT
Tropical waters worldwide; shallow subtidal zone around reefs

LIFE CYCLE
Triphasic, with isomorphic gametophyte (dioecious) and tetrasporophyte (see page 35)

Fronds are relatively small, usually growing to around 4 in (10 cm) in length, and often look bushy.

The notoriously invasive *A. spicifera* is a particular problem in the waters around the Hawaiian islands. *A. spicifera* propagates vegetatively and relatively unobtrusively: it does not form large nuisance blooms, but slowly moves into areas occupied by native species and outcompetes them.

Much effort has gone into trying to control the spread of *A. spicifera*. It is known to thrive in areas with low herbivory and is eaten by green sea turtles and various fish, so the best way to keep it down may be biological control. Interestingly, *Acanthophora* species produce a wide range of phycochemicals, which may help to make them more successful invasive species by rendering them less tasty to grazers. Phycologists are therefore exploring this chemical repertoire in the hope of developing new pharmaceuticals.

OPPOSITE | The spiny branches of *Acanthophora pacifica* (named after the Pacific Ocean), native to Hawaii and under threat from its cousin, *Acanthophora spicifera*.

ABOVE | *Acanthophora ramulosa* (Latin: "branchlike") off the coast near Perth, Western Australia, demonstrating why one common name for the *Acanthophora* genus is "spiny red seaweed."

OSMUNDARIA

Found in the Mediterranean Sea, *Osmundaria volubilis* was described by Linnaeus in 1759 as *Fucus volubilis*. However, *Osmundaria* is primarily a southern hemisphere genus and was named by Jean Vincent Lamouroux in 1813, from *O. prolifera* specimens that were perhaps sent to him by his correspondents from what was then called New Holland and is now Australia.

O. prolifera and its close relative *O. fimbriata* are subdominant in the subtidal zone in many Australian waters, as is *O. serrata* off South Africa and *O. obtusiloba* along the coastline of Brazil and the Caribbean. *Osmundaria* fronds are pleasingly distinctive. The fronds are relatively large, growing to perhaps 12 in (30 cm) long. They are branched near the base and, when fertile, reproductive cells form along the edge of the blades. The blades themselves show much variation between different species in the genus: *O. prolifera*, for example, has branches covered in feltlike hairs, whereas the blade of *O. serrata* has spines that look a little like those on a holly leaf.

RIGHT | The distinctive, corkscrew-like blades of *Osmundaria prolifera* (Latin: "spreads") growing off Cape Leeuwin on the southwest tip of Western Australia.

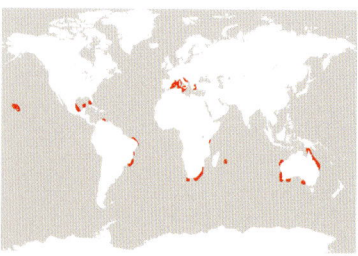

ETYMOLOGY
Possibly from the resemblance to the fern genus *Osmunda*, which was named by Linnaeus

DIVERSITY
About five to ten currently accepted species

SHAPE & SIZE
Variable but branched and often corkscrewed; up to 12 in (30 cm) long

HABITAT
Subtidal zone, often as a sub-canopy

LIFE CYCLE
Triphasic, with isomorphic gametophyte and tetrasporophyte (see page 35)

Adventitious growth, that is to say unusual and unexpected growth, can occur from older tissue: short-bladed branches can spring from the midline of older blades, especially near their bases. These new growths may look a different color and have a different younger, texture, so they were originally thought to be an epiphyte and named as the species *Vidalia gregaria*.

However, perhaps the most distinctive thing about many *Osmundaria* species is that their straplike blades are often twisted like corkscrews. The exact purpose of this helical twisting remains unclear, but it makes *Osmundaria* hard to miss and turns a simple flat blade into a more complex and three-dimensional structure.

BOSTRYCHIA

Species in the genus *Bostrychia* are small and delicate and are eaten in salads in Myanmar. They offer a good example of one of several genera that grow as epiphytes on mangrove roots in brackish waters.

Bostrychia grows in tufts on the prop roots of mangroves and, for obvious reasons, is sometimes called "sea moss." The *Bostrychia* thalli have relatively simple architectures: the polysiphonous main axis can have irregular monosiphonous beadlike branches. Fronds are usually small (shorter than around 4 in/10 cm) and feathery with curled tips. The lower branches of *Bostrychia* often carry smaller curved branchlets themselves.

ETYMOLOGY
Greek *bostrychos* = "something twisted or curled" (e.g. hair or tendril)

DIVERSITY
Around 50 species are currently known

SHAPE & SIZE
Small and tangled, hence their common name of "sea moss"

HABITAT
Epiphytic, usually on mangroves

LIFE CYCLE
Triphasic, with isomorphic gametophyte and tetrasporophyte (see page 35)

However, the morphology of *Bostrychia* thalli is very variable, which makes it hard to unambiguously identify a species from its shape. Genetic studies suggest that there is cryptic species diversity, which is where two different species look the same, and it is likely that more *Bostrychia* species exist than we currently realize.

We are also still studying population structures in this genus. Different species are found on mangrove roots at different times of year and they can use different reproductive strategies in different areas. In the French South Pacific territory of New Caledonia, for example, *Bostrychia* thalli on one side of the island reproduce through mitospores to regenerate the diploid tetrasporophyte, while on the other side of the island they reproduce sexually,

TOP | A patch of *Bostrychia tenella* (Latin: "delicate") off the coast of Lord Howe Island, New South Wales, Australia, showing why members of the *Bostrychia* genus are sometimes called "sea moss."

ABOVE | The feathery fronds and curled tips of *Bostrychia tenella*.

cycling through haploid gametophyte and diploid tetrasporophyte generations.

The best-known member of the genus *Phycodrys* is perhaps *P. rubens*, the "sea oak" described by Linnaeus in 1753 and whose oak-leaf-like blade makes its common name so very appropriate. However, the most interesting members of the genus are probably its parasites (see below).

The genus is part of the Delesseriaceae family, which is unusual in having species with veined leaves. These are not true veins but provide structural support for the blades. The blades grow from a stolon-like system, which emerges from a discoid holdfast. In *P. rubens*, the blade is monostromatic and the midrib and veins are thicker and contain more layers, but other species have slightly varied numbers of cell layers in the blade and veins. The blades grow from an apical meristem but the main axis also shows intercalary growth. When fertile, cystocarps are scattered over the blade, especially toward its edge.

The deep red *Phycodrys* members are common in the subtidal zone, where they are one of the subdominant species in the kelp forests of the northern hemisphere. *Phycodrys* species sometimes grow as epiphytes on these kelps and are, in turn, settled on by shellfish larvae. We rarely see *Phycodrys* species, however, because they can live as far as 65 ft (20 m) down, well below the low-tide mark. There is a lot of variation between individuals from the same species between the Baltic and the North seas and evidence of waves of repopulation after ice ages. *Phycodrys* species can adapt to these colder conditions by changing their fatty acid compositions.

A particularly interesting group of *Phycodrys* species are the parasites: some *Phycodrys* species are parasitized by other red seaweeds and recent molecular work has shown that these parasites are also members of the *Phycodrys* genus. *P. novae-zelandiaephila*, for example, grows as a parasite on *P. novae-zelandiae*. In effect, some *Phycodrys* species have evolved into their own parasites, which is a remarkably unusual evolutionary event and one that will keep researchers happy for years to come.

OPPOSITE, ABOVE | "Sea oak." *Phycodrys rubens* (Latin: red) growing in the Kattegat off eastern Denmark.

OPPOSITE, BELOW | A stand of *Phycodrys rubens*, showing its midrib and the characteristic veins of the Delesseriaceae family.

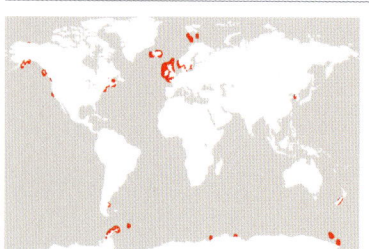

ETYMOLOGY
Greek *phycos* = "seaweed" + *drys* = "oak"

DIVERSITY
Around 25–30 currently accepted species

SHAPE & SIZE
Highly variable within the genus; larger species can be around 40 in (100 cm) long

HABITAT
Deeper subtidal zone of the northern hemisphere, often around kelp forests

LIFE CYCLE
Triphasic, with isomorphic gametophyte and tetrasporophyte (see page 35)

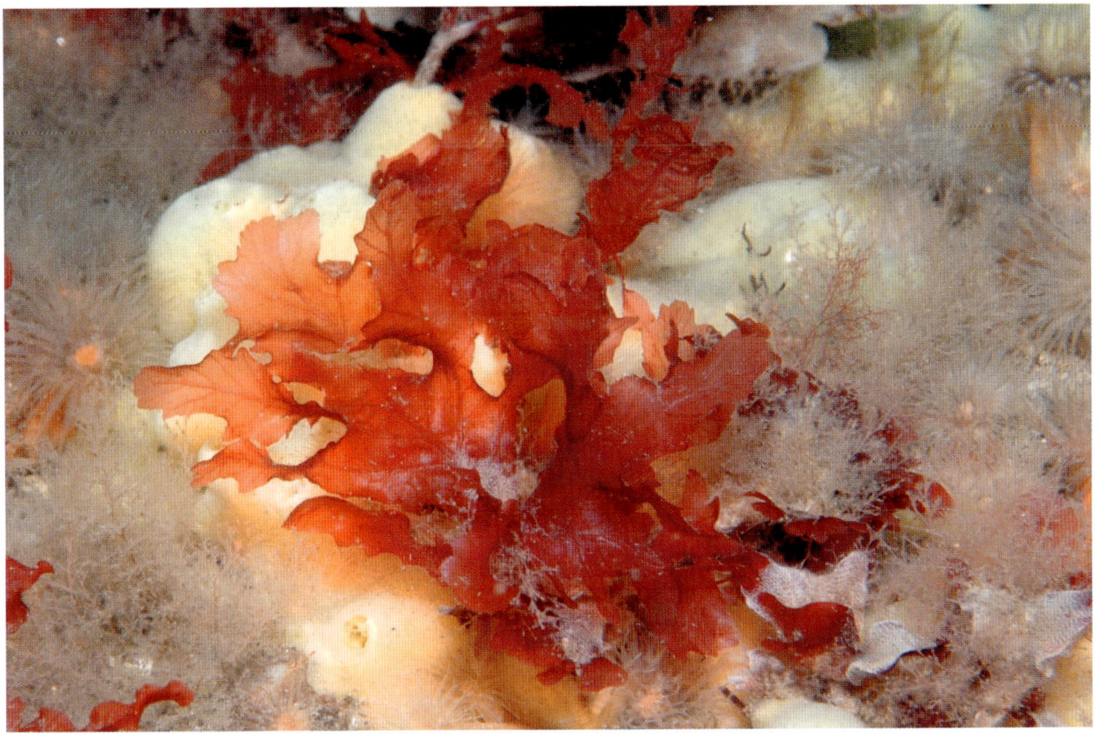

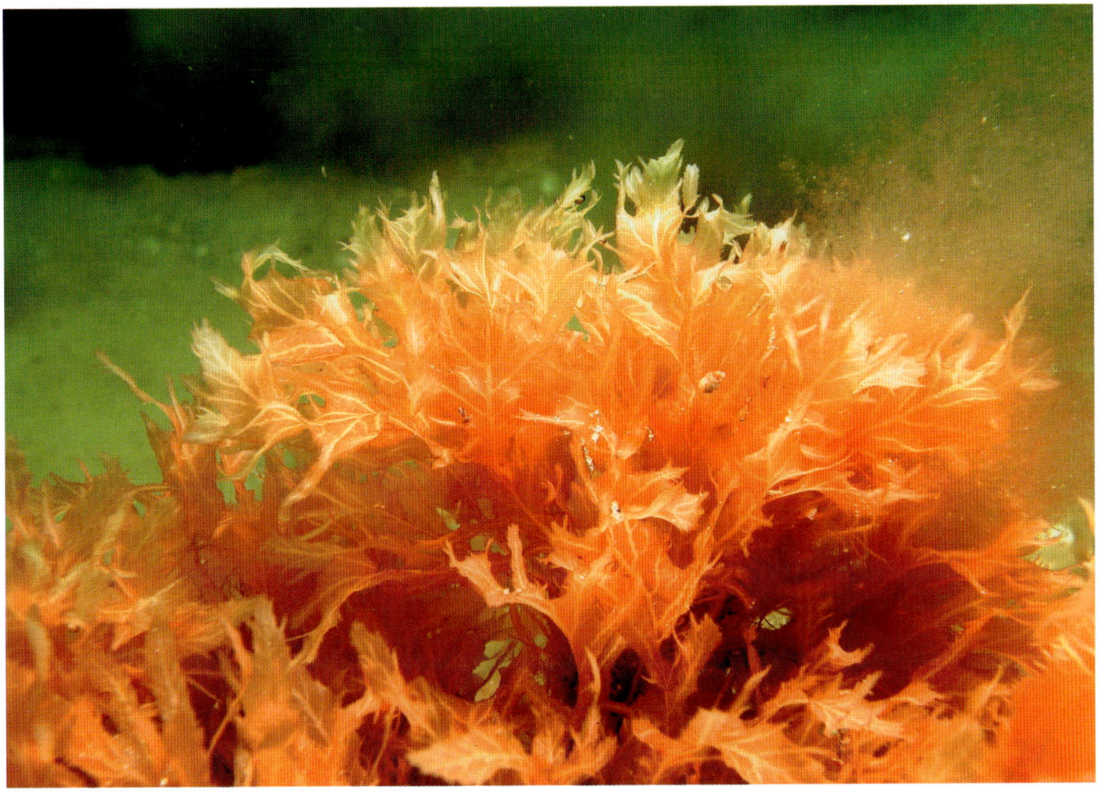

POLYNEURA

Sitting in the same family as *Phycodrys* (see page 158), *Polyneura* also has veins. The fronds have beautiful, lobed thalli with stipes that, in some species, can turn into a midrib.

There is variation within the genus, but species are often a pink or soft red color, with delicate monostromatic blades. They can also display captivating iridescence when underwater.

The thalli show irregular branching from the main axis. *Polyneura* is reasonably closely related to *Phycodrys* and grows in a similar fashion: the blade grows from an apical meristem, but the main axis also shows intercalary growth, so can elongate as the blades are growing. When fertile, reproductive spermatangia and carpogonia form in a scattered fashion, mainly in the outer parts of the blade. Many

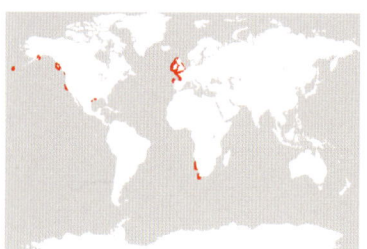

ETYMOLOGY
Greek *poly* = "many" + *neura* = "string" or "tendon"

DIVERSITY
Around half a dozen currently accepted species

SHAPE & SIZE
Distinctive, flat, veined fronds, with larger species reaching perhaps 20 in (50 cm)

HABITAT
Sheltered subtidal areas

LIFE CYCLE
Triphasic, with isomorphic gametophyte (usually dioecious) and tetrasporophyte (see page 35)

Polyneura species may also spread through vegetative propagation. Polyneura species usually grow in sheltered areas and can act as hosts for kelp gametophytes, which lodge themselves in the extracellular spaces of Polyneura and grow as endophytes in the Polyneura thallus. This tendency to support smaller seaweeds is also evident in the parasites that, as in Phycodrys, have evolved from one Polyneura species into a second species that parasitizes the first, original, species. One parasitic red species called P. latissimicola, for example, has recently been discovered to have evolved from its host, which is called P. latissima.

ABOVE AND OPPOSITE | The leafy, veined blades of *Polyneura bonnemaisonnii*, named after the French phycologist Théophile Bonnemaison (1774–1829). Other *Polyneura* species can grow as small, spiky parasites on these blades.

LEFT | The blade of *Polyneura bonnemaisonnii* under the microscope, showing the layer of cells that make up its leafy thallus.

GRIFFITHSIA

There are many species named after people, but very few genera: *Griffithsia* is one such genus. It was named in the early 1800s by Swedish botanist Carl Agardh (1785–1859), in honor of Amelia Griffiths, who was one of the great early seaweed collectors. Her collaborator William Henry Harvey wrote in the introduction to his 1849 *Manual of the British Algae*: "I am consequently wholly dependant [sic] on the liberality of my algological friends. Among these Mrs. Griffiths deservedly holds the first place."

Griffithsia thalli consist of monosiphonous axes with large cells and irregular branching. They can grow to perhaps 8 in (20 cm) long, but they are delicate because the axis is rarely more than a couple of hundred micrometers across, so thalli often end up looking like small tufts. Different *Griffithsia* species have different shaped cells, with some being

LEFT | A mat of *Griffithsia corallinoides* (Latin: "coral-like") growing in County Galway, Ireland.

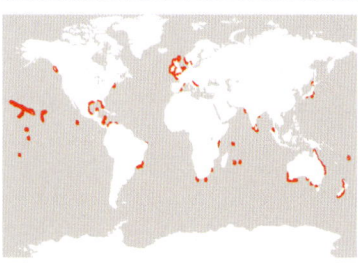

ETYMOLOGY
After Mrs Amelia W. Griffiths (1768–1858), an English phycologist

DIVERSITY
Around 30–50 currently accepted species

SHAPE & SIZE
Small and tufty

HABITAT
Widespread in the intertidal and subtidal zone, often as epiphytes

LIFE CYCLE
Triphasic, with isomorphic gametophytes (dioecious) and tetrasporophytes (see page 35)

cylindrical and others more dumbbell-shaped, but the cells are large and unusually visible to the naked eye, giving them the appearance of tiny beads on a string. The spermatangia and sporangia form around the nodes between the large siphonous cells or at the ends of the branches. *Griffithsia* species can grow on rock or as epiphytes and their holdfasts can show secondary growth, so that the base of the thallus can look rather like a banyan tree.

The delicate nature of the fronds mean that *Griffithsia* species are often eaten, especially around coastal Asia. The genus is also scientifically interesting because damaged cells may be repaired by the cells on either side, which suggests that these apparently simple structures are maintained with more care than we usually imagine seaweeds to show. This is still an area open to research, however.

ABOVE | *Griffithsia opuntioides* (Greek: "cactuslike"; under a microscope this species looks like a prickly pear), alongside the green seaweed *Ulva* and the brown *Dictyopteris*, off the coast of Galicia in northern Spain.

RIGHT | *Griffithsia monilis* (Latin: "necklace"), showing the characteristic beaded axes of the *Griffithsia* genus.

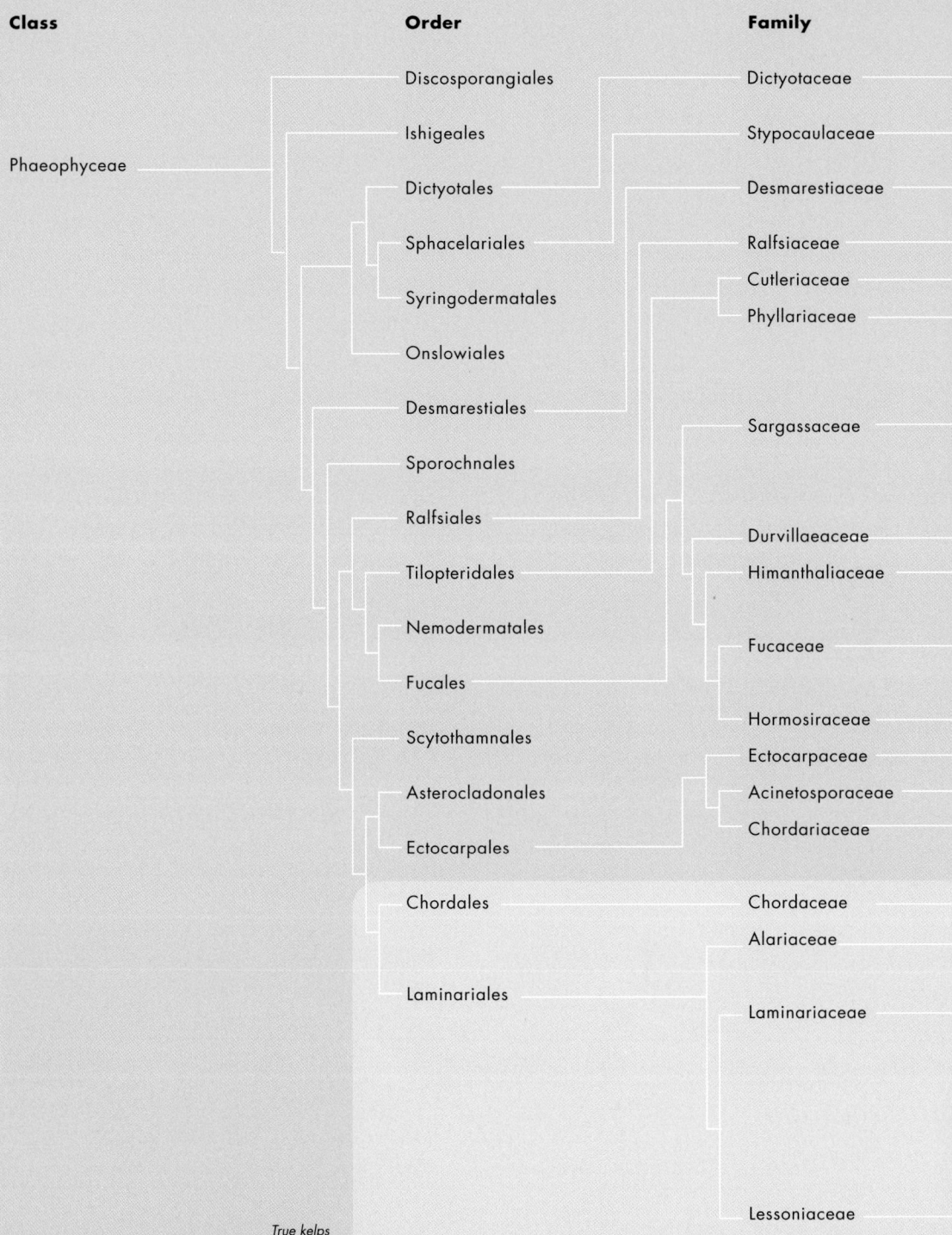

Class

Phaeophyceae

Order

Discosporangiales
Ishigeales
Dictyotales
Sphacelariales
Syringodermatales
Onslowiales
Desmarestiales
Sporochnales
Ralfsiales
Tilopteridales
Nemodermatales
Fucales
Scytothamnales
Asterocladonales
Ectocarpales
Chordales
Laminariales

Family

Dictyotaceae
Stypocaulaceae
Desmarestiaceae
Ralfsiaceae
Cutleriaceae
Phyllariaceae
Sargassaceae
Durvillaeaceae
Himanthaliaceae
Fucaceae
Hormosiraceae
Ectocarpaceae
Acinetosporaceae
Chordariaceae
Chordaceae
Alariaceae
Laminariaceae
Lessoniaceae

True kelps
(Generally speaking)

Genus

Padina

Dictyopteris

Halopteris

Himantothallus

Desmarestia

Analipus

Cutleria

Saccorhiza

Sargassum

Turbinaria

Halidrys

Cystoseira

Gongolaria

Durvillaea

Himanthalia

Ascophyllum

Fucus

Pelvetia

Hormosira

Ectocarpus

Pyliaella

Nemacystus

Chorda

Undaria

Laminaria

Saccharina

Macrocystis

Pelagophycus

Nereocystis

Postelsia

Eisenia

Ecklonia

Egregia

LEFT | A phylogenetic tree of the brown seaweeds, showing the species that feature in the following pages.

PHAEOPHYCEAE
THE BROWN SEAWEEDS

Brown seaweeds contain the wracks and kelps and dominate many of the world's intertidal and offshore subtidal zones. They are relatively young in evolutionary terms. There are no unambiguous brown seaweed fossils older than perhaps 20 MYA and DNA-based phylogenies suggest that they evolved only in the last 200–250 million years.

How the browns acquired their chloroplasts, and therefore became seaweeds, remains difficult to explain. Brown chloroplasts have many similarities to red chloroplasts, but also significant differences. It is thought that a complex web of multiple secondary, and even tertiary, endosymbioses has created several photosynthetic algal groups and that one of those lineages grew into the Phaeophyceae.

The browns have different ultrastructures to the reds and greens. Most reproduce sexually through single-celled unilocular sporangia, but some have a second, multicellular, kind of sporangium, called plurilocular. Plurilocs are generated by mitosis and each cell in the pluriloc releases a reproductive cell, most of which have two flagella: one that is smooth and one that is covered in fine hairs called mastigonemes.

In terms of biochemistry, the browns store sugars in a polymer called laminaran, which is similar to the starches of the reds and greens. Their cell walls contain characteristic polymers called alginate and fucoidan and they have evolved pigments beyond those found in the reds, notably chlorophyll c and the xanthophyll pigment fucoxanthin.

The browns are divided into several major orders. The two most prominent are the Laminariales and the Fucales. Broadly speaking, the Laminariales are kelps and the Fucales are wracks, although common names do not always map onto the scientific ones. The southern hemisphere *Durvillaea*, for example, is usually referred to as a kelp but belongs to the Fucales.

PADINA

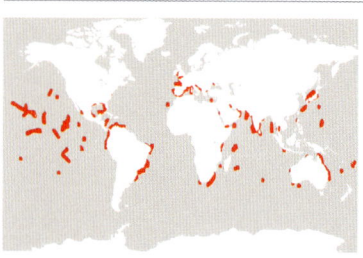

ABOVE | *Padina pavonica* (Latin: "peacock's tail") growing alongside the green *Codium tomentosum* and the bushy, brown *Ericaria tamariscifolia* off the coast of Galicia, northern Spain.

OPPOSITE | *Padina pavonica* thalli (top). Growth occurs from the rim of the thallus. (Bottom) The white aragonite that stiffens the thalli.

This diverse, warm-water genus likely appeared between the Jurassic and Cretaceous periods, some 150 MYA. Its members are extremely distinctive, which is why Linnaeus included one in his original 1753 classification of the plants. *Padina* species have fan-shaped thalli that grow in clusters

ETYMOLOGY
Greek *pedinos* = "flat"

DIVERSITY
A reasonably rich genus, with around 50 known species

SIZE & SHAPE
Small and fan-shaped, growing in clusters. Most species do not grow to more than around 6 in (15 cm)

HABITAT
Tropical and warmer waters, intertidal to subtidal zones

LIFE CYCLE
Perennial, although the thallus can regrow; some species can vegetatively propagate; isomorphic alternation of haploid gametophyte and diploid sporophytes (see page 37)

that can look a little like their common name of "peacock's tail." Growth occurs along the rim of the thallus and, to protect these delicate growing meristems, the rim curls up. Tiny hairs appear at regular growth intervals, giving the appearance of growth rings. Reproductive structures develop between these rings of hairs. The thallus itself is delicate, being only a few cell layers thick. There is a distinct cortical layer, in which the cells contain chloroplasts and are photosynthetic, and a more transparent medulla, in which cells lack chloroplasts and so do not photosynthesize.

The fragility of the *Padina* thallus is countered by one astonishing adaptation: like the green *Halimeda* (see page 86) and the red *Corallinales* (see pages 110–13), *Padina* species are able to deposit limestone—specifically, rectangular aragonite. Limestone deposition tends to happen more toward the outer parts of the thallus and *Padina* fronds can end up looking a little like small, white pottery shards lying underwater. In Indonesia, where some *Padina* species are eaten, the plants are first blanched before being ground into a Turkish delight-like paste.

DICTYOPTERIS

Like *Padina* (see page 166), *Dictyopteris* is a member of the order Dictyotales. The ancestors of the Dictyotales diverged from the ancestors of the kelps and wracks around 200 MYA, so not too long after the evolution of the brown seaweeds. This makes them quite a distinct group: their reproductive cells have only one flagellum, rather than the usual two, for example. The order itself is named after the genus *Dictyota*, which is reasonably similar to *Dictyopteris*. As with most of the Dictyotales, *Dictyopteris* species show alternation of isomorphic generations and both generations grow from a discoid holdfast, which makes it hard to tell whether any given plant is a haploid gametophyte or a diploid sporophyte. Like many of the Phaeophyceae, the thalli of *Dictyopteris* display dichotomous branching. Depending on the species, the blades may be thin and membranous (distromatic) or slightly thicker, with cortical and medullary layers. Some of the more robust species can look a little like wracks but when fertile, conceptacles appear as darker patches across the entire thallus, while in wracks they appear in receptacles at the tips of fronds. *Dictyopteris* differs from *Dictyota* in various ways, but perhaps the most notable is that *Dictyopteris* fronds

ABOVE | *Dictyopteris polypodioides* (Greek: "many-footed"), with its characteristic midrib.

have a characteristic strengthening midrib that *Dictyota* lacks.

Few of the Dictyotales have significant commercial applications, although many may be eaten. Some *Dictyopteris* species are recognized as Hawaiian *limu* and the eminent French phycologist Camille Savageau (1861–1936) was said to have

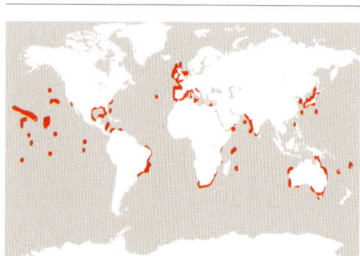

ETYMOLOGY
Greek *diktyotos* = "netlike" + *pteron* = "wing"

DIVERSITY
A reasonably diverse genus, with around 40 known species

SIZE & SHAPE
Dichotomous branches; larger individuals can be around 12 in (30 cm) long

HABITAT
Usually subtidal zones

LIFE CYCLE
Isomorphic alternation of haploid gametophyte and diploid sporophytes (see page 37)

particularly enjoyed their taste. Extracts from *Dictyopteris* species were also part of the traditional medicine of a number of Mediterranean countries, being applied to treat tuberculosis and its complications. This may not be entirely unsound: many brown algal extracts do possess some pharmacological activity, although these are still being explored.

TOP | *Dictyopteris polypodioides* growing off the coast of Galicia in northern Spain. The fronds can look brown, as they do here, or green as opposite, depending on conditions.

ABOVE | *Dictyopteris* from a distance. The species on this Hawaiian shoreline are often collected to make the local relish *limu*.

HALOPTERIS

ABOVE, TOP | *Halopteris filicina* (Latin: "threadlike") growing underwater off the coast of Sicily, Italy.

S mall and feathery, *Halopteris* species often grow as epiphytes on larger seaweeds such as *Cystoseira* (see page 192). The genus is distinctive enough that *H. scoparia* was first described (as *Conferva scoparia*) by Linnaeus in 1753. The body plans of *Halopteris* species give some insight into the

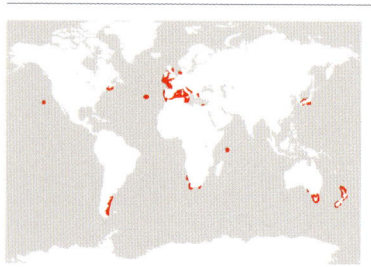

ETYMOLOGY
Greek *halos* = "the sea" + *pteron* = wing

DIVERSITY
A reasonably diverse genus, with around 40 known species

SIZE & SHAPE
Fairly small and feathery, not usually growing to more than 4 in (10 cm)

HABITAT
Subtidal zone, usually on rocks

LIFE CYCLE
Isomorphic alternation of haploid gametophyte and diploid sporophytes; a range of gamete sizes, with some species being oögamous (see page 37)

evolution of seaweed architecture. They display heterotrichy, which translates as "different hairs" and means that the body of *Halopteris* consists of horizontal filaments that lie flat along whatever surface the *Halopteris* is growing on, as well as erect filaments that grow upward from the flat ones. In *Halopteris*, one erect filament begins to grow as a single axis from an apical meristem. This filament divides lengthwise once, then again, so that the growing erect filament contains four columns of cells, giving it a cross section that looks a little like a child's drawing of a window. Alternating branches then arise that themselves develop smaller, multipinnate branchlets. Finally, oval reproductive sporangia develop in the axils (the armpits of the

branches). When viewed under a microscope, the overall effect is of a microscopic Christmas tree, hence its common name of "sea fern." The way in which this relatively regular shape can arise from a simple developmental program is something that can help us to understand the larger and more complex forms of the other browns.

DESMARESTIA

This genus gives its name to its order, the Desmarestiales, which is ecologically important in providing the dominant subtidal seaweeds in the Antarctic. The genus *Desmarestia* spans a range of branched shapes and patterns, all growing from a bulbous holdfast and, when fertile, having reproductive sporangia scattered over their thalli. Young *Desmarestia* species start as a light, olive color and darken as they get older.

Desmarestia shows alternation of heteromorphic generations and the visible fronds are the diploid sporophytes, which are perennial. Some species show alternating branches, while others show opposed branches. All show simple internal

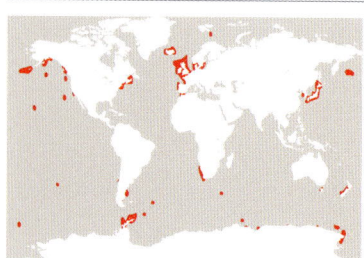

ETYMOLOGY
Named after the French zoologist Anselme Gaëtan Desmarest (1784–1838)

DIVERSITY
A reasonably diverse genus, with around 30 known species

SIZE & SHAPE
Branched, with larger species being up to 16½ ft (5 m) long

HABITAT
Lower intertidal and subtidal zones, out to around 50 ft (15 m) deep in colder waters; some species are found in the colder deep waters of the tropics

LIFE CYCLE
Heteromorphic alternation of generations, with visible diploid sporophytes and microscopic haploid gametophytes; the sporophytes can be monoecious or dioecious (see page 37)

organization, in which the thallus interior (the medulla) is made up of filamentous cells tangled together like spaghetti, rather than being made of the more neatly packed cells seen in land plants.

Their growth is trichothallic, which means that microscopic hairs form at the tips of axes and then extend. When fertile, *Desmarestia* sporophytes develop unilocular sporangia that release meiospores that grow into microscopic gametophytes. Sperm released by the male gametophytes fertilize the egg while the egg is still attached to its female gametophyte, so new visible diploid fronds grow once again from microscopic haploids.

One final and unusual adaptation worth mentioning for some *Desmarestia* species (for example, *D. ligulata*) is that their vacuoles contain

OPPOSITE | Sporophyte generation of *Desmarestia aculeata*.

TOP | Sporophyte generation of *Desmarestia ligulata*.

ABOVE | The fronds of *Desmarestia ligulata* are broad and flat.

high levels of sulfuric acid. Exactly why remains unclear, but one unfortunate side effect is that this makes these species difficult to cultivate in the same tank as other seaweeds: the sulfuric acid tends to leak out and kill the neighbors.

PHAEOPHYCEAE: DESMARESTIALES: DESMARESTIACEAE

HIMANTOTHALLUS

An usual genus, *Himantothallus* has only one known species: *Himantothallus grandifolius*, which grows large and long, ruffled blades that look very much like some of the bigger kelps. *Himantothallus* is also unusual because it is found in the Antarctic, which makes it less accessible than the kelps and harder to study. Although it looks like a kelp, the early stages of its development reveal a filamentous germling that develops the trichothallic meristems found in the Desmarestiales.

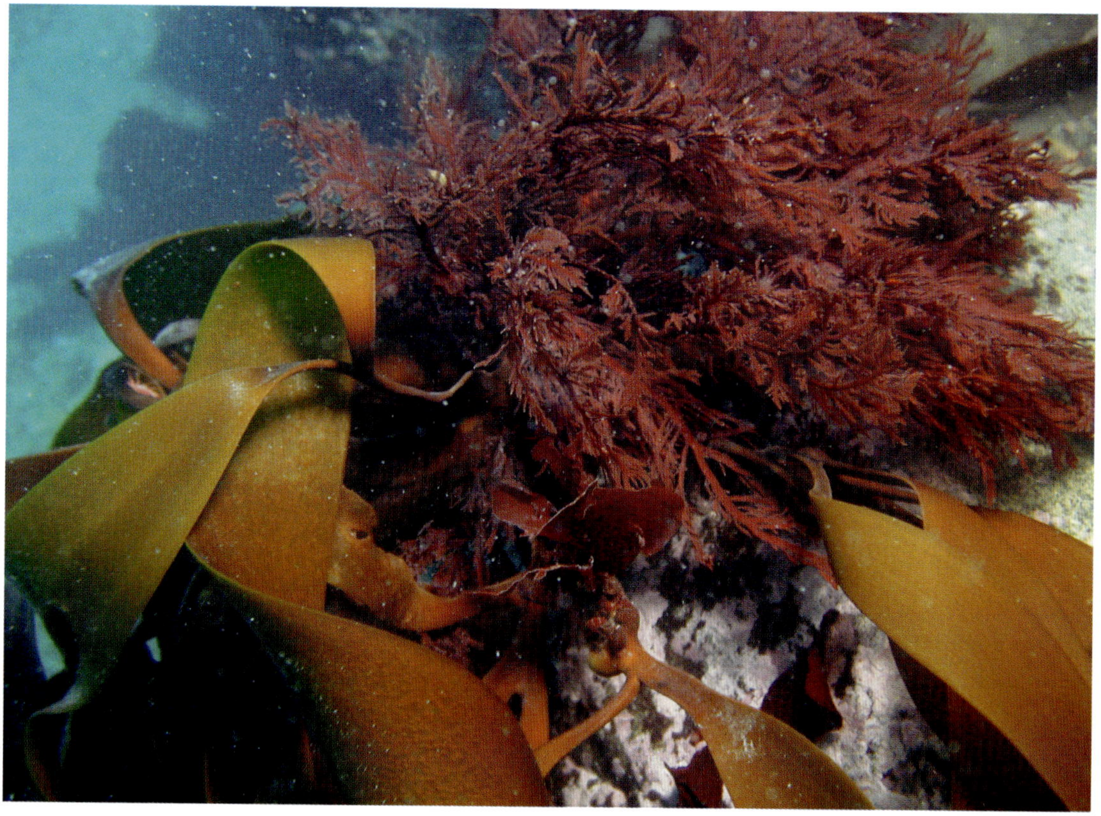

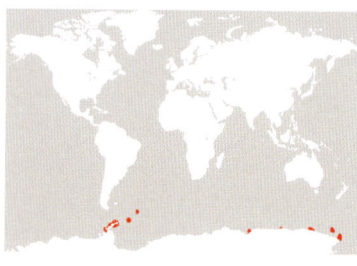

ETYMOLOGY
Greek *himantos* = "leather band" + *thallos* = "young branch"

DIVERSITY
Only one known species

SIZE & SHAPE
A large blade; larger individuals can be almost 65 ft (20 m) long

HABITAT
Subtidal zone in colder southern waters, mainly Antarctica

LIFE CYCLE
Heteromorphic alternation of generations, with visible diploid sporophytes and microscopic haploid gametophytes (see page 37)

H. grandifolius appears, therefore, to be an example of convergent evolution, in which similar traits (large blades) have arisen in distantly related orders (Desmarestiales and Laminariales) in response to similar environments (nutrient-rich offshore water).

Much about *Himantothallus* remains unknown, however. It grows fast and the tips of the thallus are colonized and eaten away by bacteria, which means that *Himantothallus* is likely to be a good source of released nutrients and probably helps to support a rich diversity of Antarctic fauna. Indeed, it is probably one of the species that used to be eaten by indigenous people in Chile, although it is difficult to be certain when looking back several centuries. *Himantothallus* shows the usual variability of the brown seaweeds in response to environmental conditions and a smoother, less ruffled form is associated with less turbulent waters. This form is different enough that it was originally identified as a separate genus, *Phyllogigas*.

OPPOSITE | The broad, blades of *Himantothallus grandifolius* (Latin: "large leaves") growing alongside the red *Georgiella confluens* (a relative of *Ceramium*) off the south coast of King George Island, Antarctica.

ABOVE | A demonstration of how large the sporophyte generation of *Himantothallus grandifolius* can be, shown here on the shore of King George Island, Antarctica.

CUTLERIA

Species of *Cutleria* show nicely heteromorphic alternation of generations. The haploid gametophytes can be fan-shaped (for example, *C. hancocki*) or attractive, dichotomously branching, membranous seaweeds (for example, *C. multifida*). When fertile, these become covered in dark spots, which are scattered reproductive sporangia. The tips of the blades are broad and blunt and look a little like ripped paper. This, however, is because they display trichothallic growth: the apparently rough edges are the tiny growing hairs. These gametophytes release gametes that, when fertilized, develop into a very different diploid sporophyte phase. The *Cutleria* sporophyte does not look much like a seaweed at all, instead appearing as a dark brown crust that grows all along its margin. When fertile, the sporophyte develops unilocular sporangia across its surface and releases gametes that, as in most brown seaweeds, find each other by chemoattraction.

The sporophyte phase of *Cutleria* looks so different that it was originally believed to be a completely different genus named *Aglaozonia*. The sporophyte is still called the Aglaozonia phase as a nod to that history (see page 32).

ABOVE | Trichothallic growth on the tips of *Cutleria multifida*.

OPPOSITE, TOP | *Cutleria kraftii*, named after the extant Australian phycologist Gerald Kraft and showing the hairs that fringe the ends of the blades.

OPPOSITE, BOTTOM | A leafy seaweed salad of the green *Ulva* and the browner blades of *Cutleria adspersa* (Latin: "sprinkled").

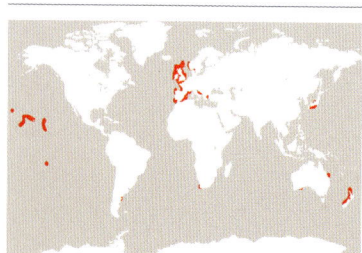

ETYMOLOGY
Named after Catherine Cutler (1784–1866), a nineteenth-century British phycologist

DIVERSITY
A reasonably diverse genus, with around 15 known species

SIZE & SHAPE
The two phases are very different shapes; one forms a crust; the other forms dichotomous branching fronds that can be around 20 in (50 cm) long

HABITAT
Usually prefers more mobile surfaces, such as sand and shingle, rather than rocks

LIFE CYCLE
Heteromorphic alternation of generations, with visible diploid sporophytes and microscopic haploid gametophytes; anisogamous (see page 37)

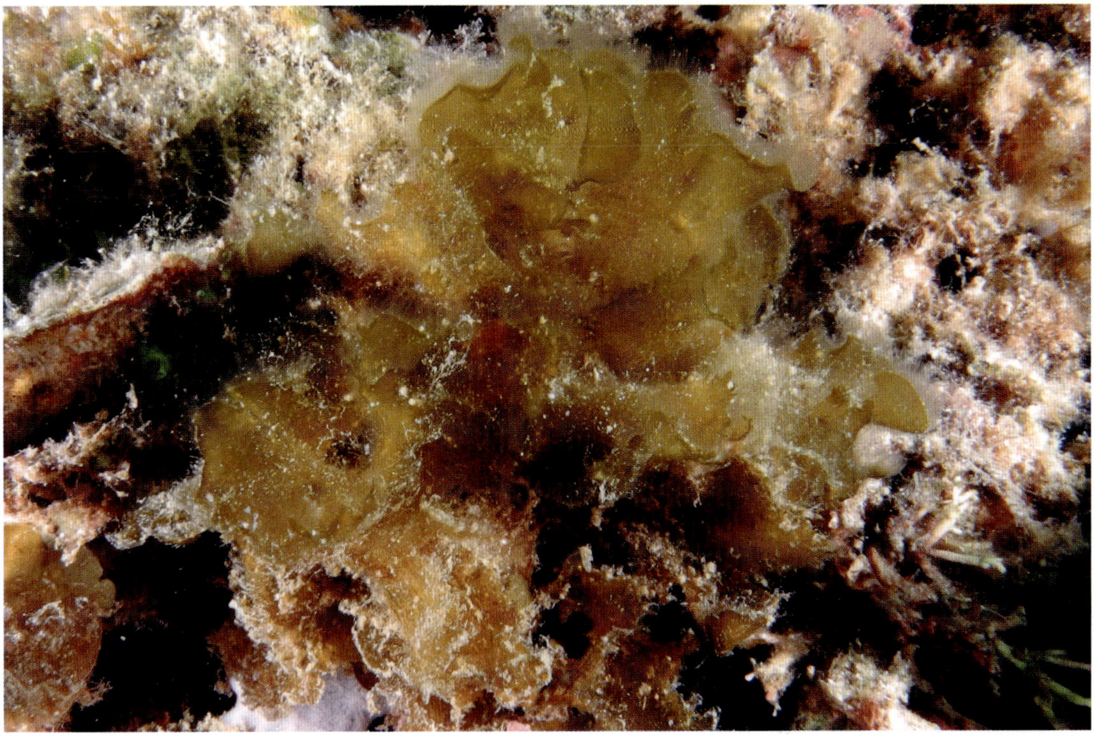

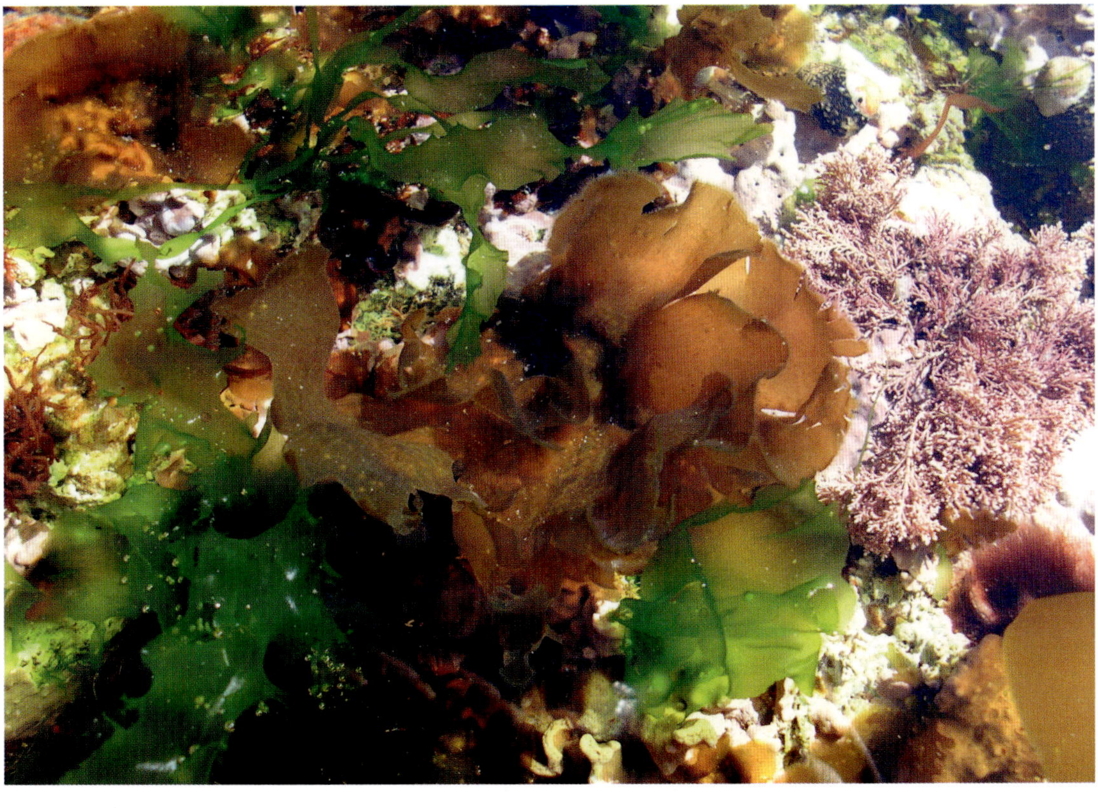

SACCORHIZA

In Georgette Heyer's novel, *Frederica*, Felix Merriville, the youngest brother of the eponymous heroine, finds his sisters immersed in "frills and furbelows." This English phrase means "unimportant things," but is a little tautologous, because a furbelow is just another word for a frill. Looking at *Saccorhiza polyschides*, therefore, it's immediately apparent why its common name is "furbelows." A beltlike stipe grows from a large, bulbous holdfast and, near the holdfast, the stipe is extravagantly frilled. The stipe develops into a large-bladed frond, which is often frayed toward its tip from wave action and erosion. *Saccorhiza* plants are fast-growing annuals that can reach a good size. They are also opportunistic and can rapidly settle on recently disturbed or scoured surfaces. All of these factors make them reasonably easy to find and harvest around the coasts of northwest Europe, so *Saccorhiza* was historically used as a common species in the production of kelp.

Given its large-bladed fronds, *Saccorhiza* was believed to be a member of the Laminariales for the better part of two centuries. Surprisingly, it is not. Along with the kelplike genus *Phyllariopsis*, *Saccorhiza* is actually one of only two large-bladed genera in the order Tilopteridales. Most of the

LEFT | The elaborately folded base of *Saccorhiza polyschides* (Greek: "many clefts"), growing here in the west of County Galway, Ireland.

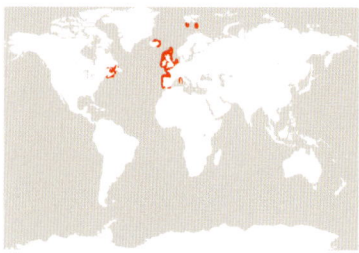

ETYMOLOGY
Greek *sacco* = "sac" + Greek *rhiza* = "root"

DIVERSITY
Only two species are currently known

SIZE & SHAPE
A large-bladed frond with a strap- or bladelike stipe; the stipe is frilled near the large, bulbous holdfast; larger specimens can be around 6½ ft (2 m) long

HABITAT
Subtidal zone, northeast Atlantic coastline and into the Mediterranean, down to around 115 ft (35 m) deep; it usually grows on rocks but can grow on less stable substrate

LIFE CYCLE
Heteromorphic alternation of generations, with visible diploid sporophytes and microscopic haploid gametophytes (see page 37)

Tilopteridales are much smaller and more delicate (for example, *Cutleria*, see page 176), but an increasing weight of ultrastructural evidence led to *Saccorhiza* being regarded as problematic and it was moved out of the Laminariales and into the Tilopteridales in 2001.

Like all the Tilopteridales, *Saccorhiza* shows alternation of heteromorphic generations and it has a place in phycological history as the species in which, in 1915, the French phycologist Camille Savageau discovered microscopic gametophytes. This established the idea that kelplike seaweeds could reproduce by an alternation of very heteromorphic generations.

ANALIPUS

The *Analipus* genus represents the order Ralfsiales and, along with *Cutleria* (see page 176), is an example of a crustose brown seaweed.

Most genera in *Analipus'* order, such as the eponymous *Ralfsia* itself, grow on rocks as spreading brown crusts in both their haploid gametophyte and diploid sporophyte generations (that is, the generations are isomorphic). *Analipus* starts as a prostrate crust of filaments but then develops a number of erect branched axes that give it the common name of "sea fir." The erect axes are seasonal, but the crusts are perennial and allow the plant to overwinter. The small, erect branches are several cell layers thick, with an outer cortical layer and an inner, nonphotosynthetic, medullary layer. They are most obvious in spring, which is when the Japanese species, *A. japonicus*, is collected and prepared as *matsumo*, which is used to flavor soups.

Unlike *Cutleria*, the crust and erect axes of *Analipus* are both parts of the same life cycle generation. The haploid and diploid generations of *Analipus* have the same morphology (they are isomorphic and crustose in both generations). In *Cutleria*, on the other hand, the crust is a separate life cycle generation to the leafy form, so *Cutleria's* two generations are heteromorphic and *Cutleria* is crustose in only one generation.

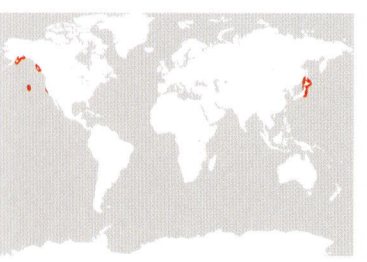

ETYMOLOGY
Greek *analipos* = "barefoot"

DIVERSITY
Only three species are currently known

SIZE & SHAPE
Small, hand-sized plants, growing from a perennial crust

HABITAT
Mid- to low intertidal zone on both sides of the North Pacific, usually on rocky substrates

LIFE CYCLE
Isomorphic alternation of haploid gametophytes and diploid sporophytes; far more *Analipus* gametophytes than sporophytes have been observed in the wild, leading to suggestions that gametes may be able to regenerate gametophytes, rather than fusing to form sporophytes (see page 37)

ABOVE | The tufty fronds of *Analipus japonicus* (Latin: "Japanese") are to the top right, struggling to make themselves seen against the distinctive sacs of the red *Halosaccion* and the broader, flatter blades of various brown species.

LEFT | "Sea fir." The seasonal part of the *Analipus japonicus* frond, which grows from a perennial crust (not shown here).

PHAEOPHYCEAE: ECTOCARPALES: ACINETOSPORACEAE AND ECTOCARPACEAE

PYLAIELLA AND ECTOCARPUS

The seaweed *Pylaiella littoralis*, or "sea felt," was the first filamentous brown seaweed described by Linnaeus in 1753, under the name *Conferva littoralis*. Filamentous browns are common, often growing as epiphytes on other seaweeds. They have no particular cultural or economic use, although they give rise to nuisance blooms and can biofoul wharves and ships' hulls. They are, however, of scientific importance because they grow well under laboratory conditions. For that reason, we have used them for the better part of a century to investigate aspects of brown seaweed biology.

Like many filamentous browns, *Pylaiella* is a member of the order Ectocarpales, which is named

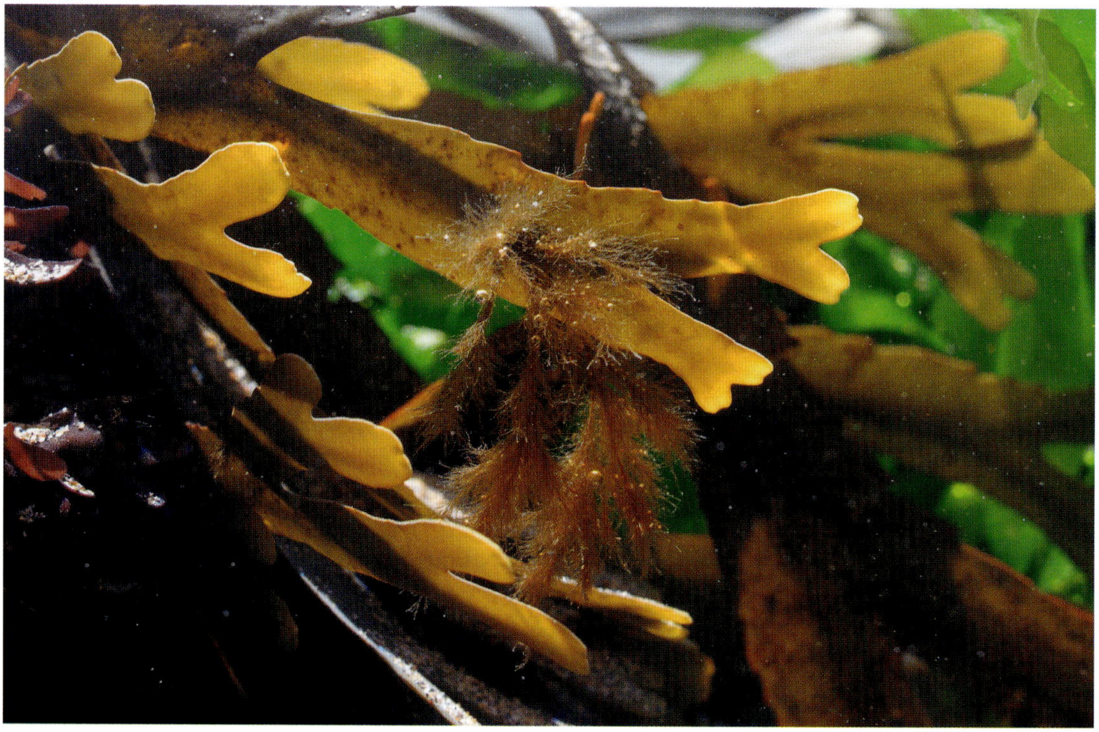

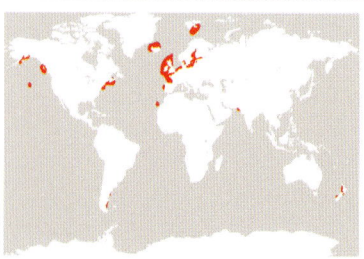

ETYMOLOGY
Named after Jean Marie de La Pylaie (1786–1856), a French explorer and botanist. The mangled anglicization of the name means that the genus used to be spelled *Pilayella*

DIVERSITY
Species are hard to delineate, but around 10–30 species are currently suggested

SIZE & SHAPE
Heterotrichous growth into a tangled mat of filaments, with larger specimens reaching 20 in (50 cm) acrosss

HABITAT
Often epiphytic but can grow attached to rocks or stones or shingle

LIFE CYCLE
Complex, based around sexual alternation of isomorphic haploid gametophytes and diploid sporophytes (see page 37)

after the genus *Ectocarpus*. The various genera and species of the Ectocarpales are almost impossible to tell apart unless you have access to a good microscope or a DNA sequencing machine.

The life cycles of the Ectocarpales are complex, but are based around sexual alternation of isomorphic haploid gametophytes and diploid sporophytes. Thalli are filamentous and heterotrichous, often branching to grow into hairy, tangled tufts. More compact *Pylaiella* tufts can develop internal architectures that begin to resemble those of the kelps, which hints at the evolution of larger thalli. Species can reproduce sexually (gametes are produced by mitosis in the plurilocs of haploid gametophytes and fuse into diploid sporophytes, which then produce meiospores by meiosis in their unilocs) or asexually (through diploid mitospores produced by mitosis in the plurilocs of diploid sporophytes) or vegetatively. Unusual life cycle loops are frequently seen (for example, unfertilized gametes may develop by parthenogenesis into either gametophytes or sporophytes). In *Pylaiella*, sporangia form in line with the other cells in the filament, so that they appear as beads on a necklace, while in *Ectocarpus* the sporangia branch off the filament.

OPPOSITE | *Pylaiella littoralis* (Latin: "seashore") growing as an epiphyte on the flat, forked fronds of its brown relative, *Fucus serratus*.

TOP | "Sea felt." Mats of *Pylaiella littoralis* can often be 12 in (30 cm) or so in length, as here, off the coast of Normandy, France.

RIGHT | Species of *Pylaiella*, *Ectocarpus*, and many of the other Ectocarpales grow as epiphytes.

NEMACYSTUS

Closely related to the *Cladosiphon* genus, *Nemacystus* is a "jelly weed." Light or golden-brown in color, jelly weeds display alternation of isomorphic generations. The thalli of both generations are built from multiple cylindrical branched axes. Depending on the species, thalli can be uniaxial or multiaxial. Each axis is several cell layers thick, with outer cortical layers and an inner medullary layer. When fertile, unilocular and plurilocular sporangia arise from the cortical layer. The inner medulla is filamentous, rather than consisting of closely packed, more spherical cells.

There is great morphological diversity across the jelly weeds, not least because their growth can be monopodial or sympodial. Monopodial growth comes from one meristem. In sympodial growth, on the other hand, the growing meristem stops and a brand-new meristem starts somewhere else on the thallus. This results in very irregular and diverse shapes across a species. This is all interesting because the body plan of *Nemacystus* is more complex than those of its close *Pylaiella* relatives (see page 182), but less complex than those of its more distant kelp cousins. We can use these differences to think about the evolution of body plans in the brown seaweeds.

Their swollen, cylindrical architecture also makes jelly weeds a tempting foodstuff and they are often cultivated and eaten in Japan as various forms of *mozuku*, a kind of noodle. Because jelly weeds are rich in sulfated polysaccharides, and since their cultivation is well established, they are a promising crop species for biotechnological purposes and, to this end, a number of draft jelly weed genomes have recently been sequenced. These are now being studied for further insights into their biology.

BELOW | A bowl of *mozuku*, made from fronds of *Nemacystus*.

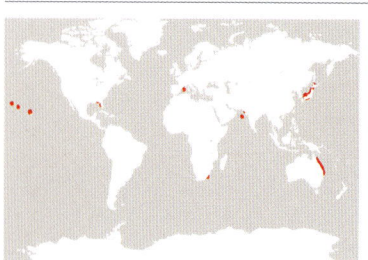

ETYMOLOGY
Greek *nema* = "thread" + *kystis* = "pouch"

DIVERSITY
Only around ten species are currently recognized

SIZE & SHAPE
Fairly small, with some individuals reaching around 8 in (20 cm) in length; cylindrical, noodlelike tubes

HABITAT
Intertidal zone

LIFE CYCLE
Isomorphic alternation of haploid gametophytes and diploid sporophytes (see page 37)

ABOVE | The April *mozuku* harvest off the south coast of Okinawa, Japan.

LEFT | "Jelly weed." The noodlelike fronds of *Nemacystus decipiens* (Latin: "deceptive," because it can be confused with other species).

PHAEOPHYCEAE: FUCALES: SARGASSACEAE

SARGASSUM

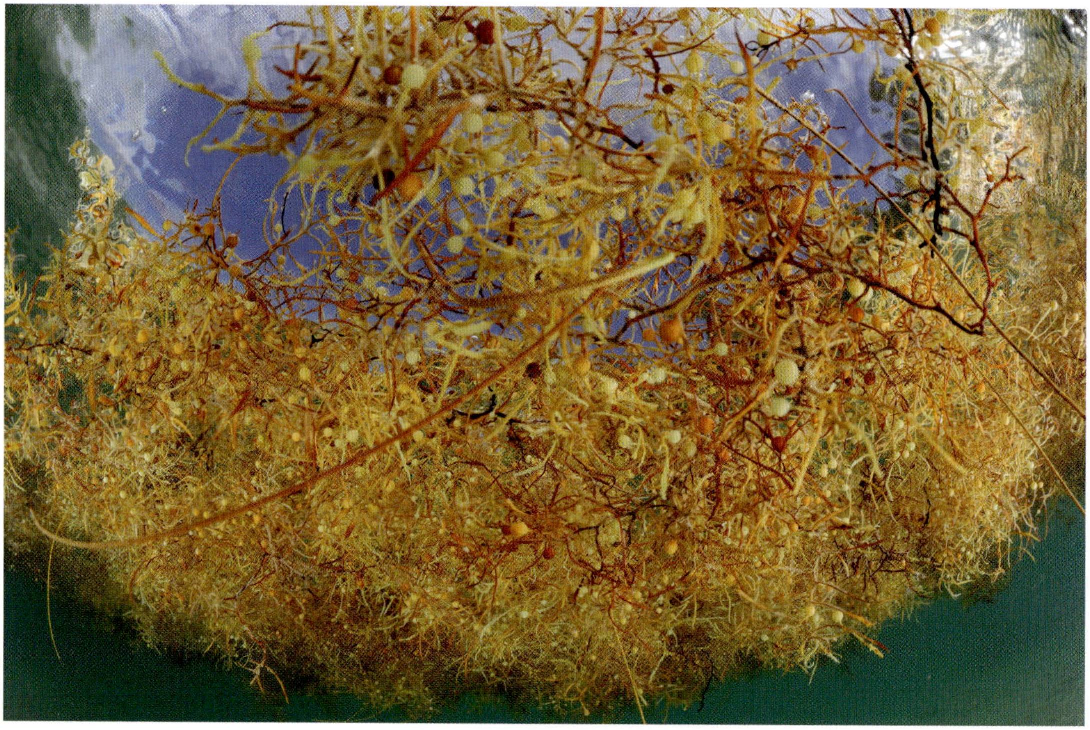

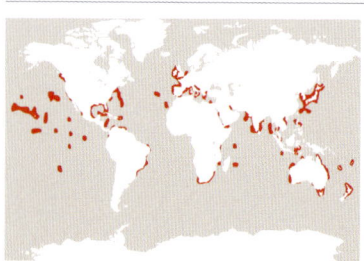

In 1778, the Swedish naturalist Carl Peter Thunberg wrote: "We sailed now in the sea called the Cross Sea, which is pretty thickly covered with Sargazo. This seaweed floats upon the surface of the ocean in incredible quantities. Sometimes it quite hides the face of the water in calm weather, so that one seems to be sailing through a meadow: at other times it forms large floating islands." The Cross Sea is today called the Sargasso Sea because of this floating *Sargassum*, and Thunberg highlights three important points about this genus. First, many *Sargassum* species grow attached to their substrate, but two species—*S. natans* and *S. fluitans*—are free-floating and can reproduce

ETYMOLOGY
Spanish *sargazo* or Portuguese *sargaço* = "seaweed"

DIVERSITY
Contains several hundred species

SIZE & SHAPE
Bushy, usually with irregular branches. Some species have air bladders to help the thallus float; larger species can be over 6½ ft (2 m) long

HABITAT
Spans the intertidal zone; most species are attached, but two are free-floating; increasingly worldwide, *Sargassum* species are invasive in a number of countries

LIFE CYCLE
Visible diploid adults produce haploid gametes (see page 36)

and grow far out to sea. Second, this oceanic growth makes rafts of *Sargassum* a haven for a range of marine species in the mid-Atlantic Ocean. Third, and increasingly problematically, *Sargassum* is an unusual genus because some of its species grow well in warmer, more tropical waters, while most brown seaweeds grow in cooler temperate waters. A combination of climate change and increased marine pollution now appears to be driving *Sargassum* invasions and harmful blooms. Many golden-brown *Sargassum* species have spread to new waters over the past few decades, and massive annual "brown tides" in the waters around the Caribbean have become a common sight, clogging beaches and smothering marine life.

OPPOSITE | Floating fronds of *Sargassum natans* (Latin: "swimming") in the Sargasso Sea off Bermuda.

TOP | Rotting *Sargassum* on Mexico's Caribbean shoreline. The smell can be overpowering.

ABOVE | A brown tide—a thick mat of rotting *Sargassum*—south of Cancún, Quintana Roo, Mexico.

HALIDRYS

The best-known member of the *Halidrys* genera is *Halidrys siliquosa*, a common perennial species once used as feedstock for nineteenth-century potash industries. It shares its common name of "sea oak" with species in several other genera (see *Phycodrys*, page 158) but, in the case of *Halidrys*, it takes some imagination to see the similarity. The thallus, which like all the Fucales is the diploid adult, is alternately branched into compressed, dark olive fronds and the stipe is attached to a conical holdfast. Some branches can develop into air bladders, while others may terminate in a reproductive receptacle. The air bladders and receptacles look reasonably similar, but an easy way to tell them apart is to remember that the air bladders are divided into internal compartments by walls that run across the width of the bladder. The edges of these walls appear as slight

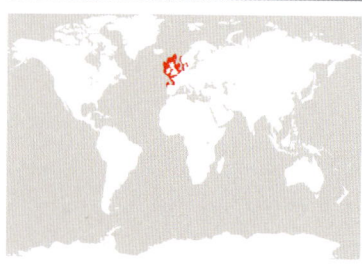

ETYMOLOGY
Greek *halo* = "sea" + *drys* = "oak"

DIVERSITY
Only two known species

SIZE & SHAPE
Can grow to around 6½ ft (2 m) long, but large specimens are usually around 40 in (100 cm)

HABITAT
Shallow subtidal zone along the eastern Atlantic coastline from Portugal up to Norway and the Baltic Sea

LIFE CYCLE
Visible diploid adults produce haploid gametes (see page 36)

depressions in the bladder, so the bladder looks a little like an elongated griddle. The receptacles, on the other hand, have no dividing marks.

As a common, moderately sized, and hardy seaweed, *Halidrys* often provides a home for a range of epiphytes. This is not a great way to live and *Halidrys* accordingly produces a range of antifouling compounds. None of these has yet found any particular market penetration, but they do give some insight into the wealth of biochemical novelty to be found in the brown seaweeds.

OPPOSITE | *Halidrys siliquosa* (Latin: "slender pods") sheltering a male short-snouted seahorse off the south coast of Devon, UK.

BELOW | The diploid adult generation of *Halidrys siliquosa*. The visible air bladders help *Halidrys* rise above other seaweeds.

PHAEOPHYCEAE: FUCALES: SARGASSACEAE

TURBINARIA

Members of this distinctive genus resemble thistle bushes and are sometimes called the "spiny leaf" seaweeds. A stiff stipe supports tough conical blades that are fringed with a double row of sharp spines. The holdfast often has multiple rhizoids. As with all Fucales, the visible thallus is the diploid adult and there is no visible haploid generation, only haploid gametes. *Turbinaria* blades are perennial and usually a yellow-brown color, but can be darker. Despite the stiffness of its fronds, *Turbinaria* used to be collected and used as a relish in the Riao archipelago to the south of Singapore.

Turbinaria species often grow as single individuals but can form clumps or thick mats. As its appearance

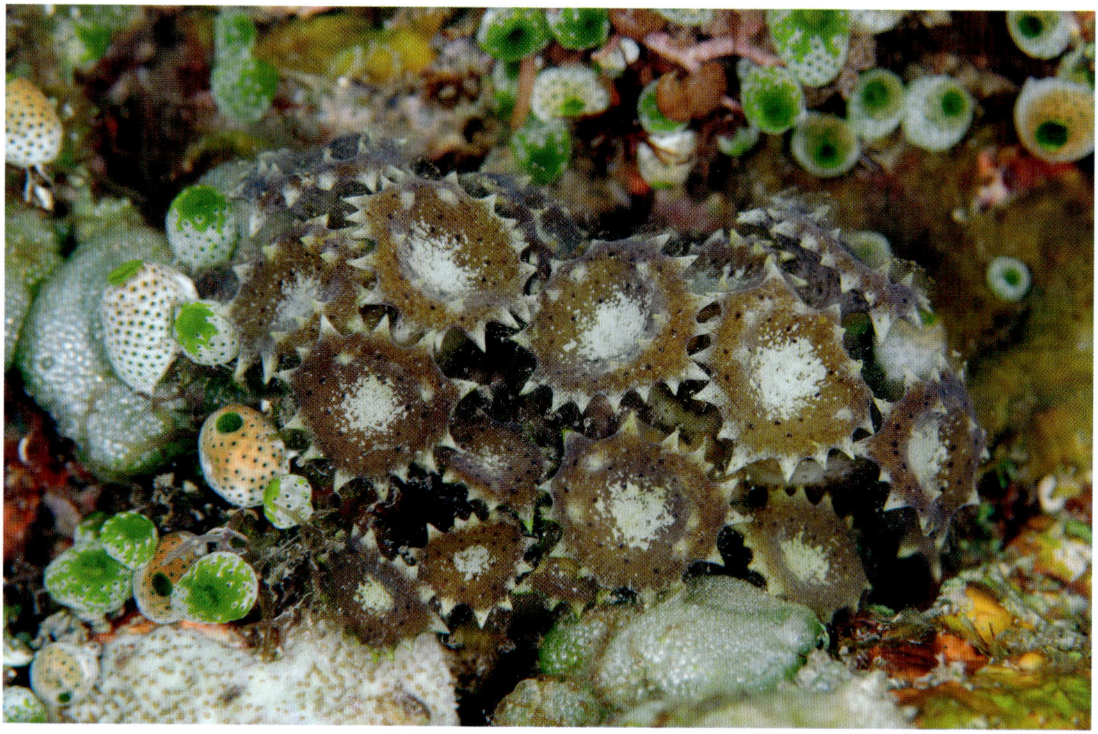

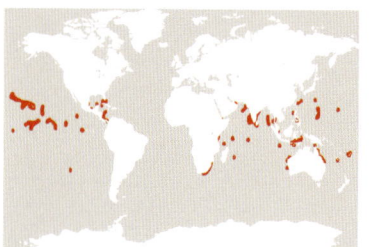

ETYMOLOGY
Latin *turbin* = "that which spins or twirls around"

DIVERSITY
Around 20 species are known, with a few more uncertain

SIZE
Larger species can reach around 12 in (30 cm)

HABITAT
Widespread in tropical and subtropical regions. Found in the mid-intertidal to subtidal zones around 100 ft (30 m) deep. Will grow on a range of substrates, including on coral reefs

LIFE CYCLE
Sexual or vegetative; can be monoecious or dioecious; oögamous; visible diploid adults produce haploid gametes (see page 36)

would suggest, this is a tough genus. Its members can withstand strong wave action and the fronds can regrow from the holdfast if they have been scoured away. To add to this, they can grow by vegetative propagation and their growth is sufficiently adaptable that they can sometimes develop air bladders that help scoured fragments to float. Unfortunately, while these adaptions help *Turbinaria* to survive harsh environments, they also make many *Turbinaria* species strongly invasive and the Pacific *T. ornata* is a particular nuisance. This propensity for invasion has been turned to some advantage by treating *Turbinaria* as an ecosystem indicator, especially around coral reefs. If the distinctive *Turbinaria* plants are seen to increase, this may be a sign that the ecosystem is in danger.

CYSTOSEIRA AND ERICARIA

These two genera, along with *Gongolaria* (see page 194), contain a number of species commonly called the bushy wracks. In *Cystoseira* and *Ericaria*, individuals grow together in clumps. They have perennial bodies with a range of modified branches that, for seaweeds, show considerable differentiation into different cell types. The holdfast is conical and supports a tough stipe. The thallus grows as a principal, branched, axis that is divided into a lower and upper region. Branches on the lower part of the thallus are short and flat and look like plant leaves. The upper branches in some species can swell into either storage chambers called tophules or into air bladders. These swellings are usually strung together, like peas in a pod, which is why the genus earns its name.

A delightful adaptation in many bushy wracks (for example, *E. selaginoides*, or "rainbow wrack") is underwater iridescence. Recent research shows that this is caused in a similar way to the iridescence seen on the surface of puddles of oil. The *Ericaria* can make small droplets of oil inside its cells. By bringing these droplets together, the plants form an internal oil slick that scatters and reflects light. Exactly how

ABOVE | The iridescent fronds of *Ericaria selaginoides* (Latin: "mosslike"), which until recently was called *Cystoseira tamariscifolia*.

OPPOSITE | A stand of bushy *Gongolaria baccata* (Latin: "having berries").

and why bushy wracks trigger this iridescence is still unknown, although they may be using the coalesced oil droplets as a lens to focus more light on the chloroplasts during times when light is limiting.

ETYMOLOGY
Greek *kystis* = "pouch" + *seiria* = "chain"
Greek *ereika* = "heather"

DIVERSITY
Around 40–50 known species, with another 20–30 uncertain

SIZE & SHAPE
A wide variety of forms, all of which are bushy and branched; fronds are perennial and usually less than 40 in (100 cm) long

HABITAT
Rocky intertidal zone worldwide; they can often serve as hosts for epiphytes

LIFE CYCLE
Monoecious; visible diploid adults produce haploid gametes (see page 36)

GONGOLARIA

Several species in the *Gongolaria* genus have recently been moved across from the *Cystoseira* genus. The differences between the two genera are small and involved: the cortical cells that form the outer layer of the thallus are a slightly different shape to those in *Cystoseira*, species of *Gongolaria* can be a little larger, they can have smooth primary branches in the basal region of the main axis, and they grow as very bushy plants from a single axis, rather than in the clumps seen in *Cystoseira*.

This taxonomic reevaluation is interesting and exciting. In part, that is because these fine microscopic differences shed light on the evolution of body plans in the seaweeds. But, in part, it is also because many of the species that were transferred from *Cystoseira* to *Gongolaria* were from the Eastern Mediterranean. In short, there is more biological diversity in the Eastern Mediterranean than we had realized, which is a gentle reminder that we do not understand why new species are being formed or how other marine species interact with the various bushy wracks.

ABOVE | A close-up of *Gongolaria baccata*, showing the air bladders that give the species its name.

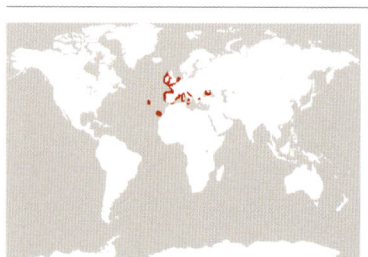

ETYMOLOGY
Italian *gongola* = "shells"

DIVERSITY
Around a dozen known species

SIZE & SHAPE
Moderate, with larger individuals reaching around 5 ft (1.5 m); plants are bushy, growing from a single main axis

HABITAT
Rocky intertidal

LIFE CYCLE
Visible diploid adults produce haploid gametes (see page 36)

DURVILLAEA

Species of the *Durvillaea* genus include the famous "bull kelps" of the southern hemisphere and, for once, the common name is appropriate. *Durvillaea* starts from a large, sturdy holdfast that grows a thick stipe and a broad, long blade. However, like *Himantothallus* (see page 174) and *Saccorhiza* (see page 178), *Durvillaea* do not sit among the "true kelps" of the Laminariales. The blades of *Durvillaea* show diffuse growth and grow fast, with prime specimens reaching over

BELOW | *Durvillaea antarctica* (Latin: "southern"), growing off the southwest coast of New Zealand.

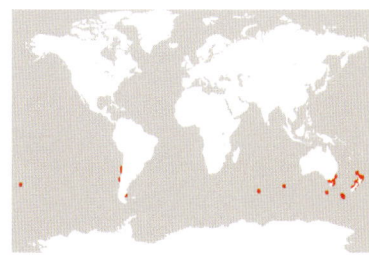

ETYMOLOGY
Named after the French naval explorer Jules Sébastien César Dumont d'Urville (1790–1842)

DIVERSITY
Five to ten known species

SIZE & SHAPE
Large, straplike blades with a honeycomb-like interior; large individuals can be 33 ft (10 m) long and have wet weights of more than 110 lb (50 kg)

HABITAT
Low intertidal to subtidal zone in the cold, exposed waters of the Southern Ocean

LIFE CYCLE
Visible diploid adults produce haploid gametes (see page 36)

33 ft (10 m) in length. The stipe can branch to give multiple blades coming off the same holdfast. When fertile, conceptacles develop, scattered along the blade.

Durvillaea competes for living space with kelps in the *Lessonia* genus along the exposed coastlines of the antipodes and its construction is suitably sturdy. The scouring of the blades will often rip them open into long straps and the blades can develop a tough, honeycomb-like interior. Famously, this structure has been put to good use as far apart as Chile and New Zealand. In Chile, indigenous peoples would cut a blade open much as we would a pita bread and use the honeycombed pouch of *D. antarctica* as a water bottle. In New Zealand, indigenous peoples would perform the same operation, but this time use the pouches as a game bag to help preserve the fat, meaty seabirds that they caught.

D. antarctica's dominance of many New Zealand intertidal regions also saw its use in traditional medicine. The Maoris call it *rimuroa* and roasted and ate it as a treatment for parasitic intestinal worms and skin complaints and fermented it for use as an ingredient in other medicines. It was even burned for use as ink by early European settlers.

Durvillaea species have a high alginate content and reasonable levels of the sugar mannitol, so there is some expectation that, in the future, they may become a viable crop in the waters of the southern hemisphere.

ABOVE | "Bull kelp." Large diploid adult *Durvillaea* on Cathedral Caves beach, southern Otago, New Zealand, showing the leathery blades used by the Maori and others.

OPPOSITE | A large blade in an exposed location needs a strong holdfast and every *Durvillaea* frond has one.

HIMANTHALIA

Linnaeus' straplike *Fucus elongatus* of 1753 was transferred into the lonely *Himanthalia* genus by the British naturalist Samuel Gray in 1821. There are only two known species in the genus. Studies have confirmed the genus' singular nature: electron microscopy, for example, has shown that *Himanthalia* sperm have unique and distinctive barbs on the tinsel flagellum of their sperm.

The species is distinctive. Its long, dark olive blades give it its common name of "thong weed," and *H. elongata* has long been one of the major species that make up the driftweeds of northern France. These were collected and used first as fertilizer for the famous artichokes of Brittany and the barley fields of the Île de Ré, off La Rochelle, and then as an occasional species for potash production. They are still sometimes pickled and eaten as a relish along the coastlines of northwest Europe, in the United Kingdom, Ireland, and France.

Himanthalia is usually biennial, spending its first year as a small, light green, button-like thallus ("button-weed") that overwinters and matures the following spring to develop into two straplike blades that display the dichotomous branching form of the Fucales. Counterintuitively, the straplike blades are not the main frond of the thallus, but are actually the reproductive receptacles. The basal frond is the button. The straplike receptacles become fertile and release their gametes in summer, before decaying. After fertilization, the zygotes settle rapidly and attach themselves to the substratum, ready to grow into buttons and start the biennial cycle once again.

ABOVE | The thong weed form of *Himanthalia elongata*. These straplike fronds are the reproductive receptacles that grow in spring.

OPPOSITE | The button-weed form. The buttons produce straplike receptacles when they become fertile.

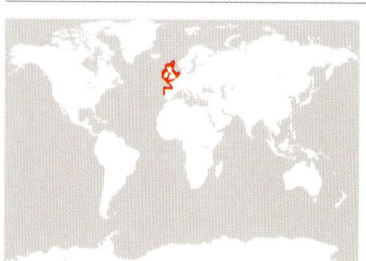

ETYMOLOGY
Greek *himan* = "leather strap" + *thalia* = "blossom"

DIVERSITY
Only a couple of known species

SIZE & SHAPE
The thallus is small and button shaped; the reproductive receptacles are straplike and can grow to around 6½ ft (2 m) long

HABITAT
Lower intertidal zone on moderately exposed shores around northwest Europe. It tends to sit between *Fucus* and the kelps

LIFE CYCLE
Visible diploid adults produce haploid gametes (see page 36)

ASCOPHYLLUM

In 1759, Linnaeus recognized *Ascophyllum nodosum* as distinct from other brown seaweeds, with the genus containing just one species, *A. nodosum*. The holdfast is disk-shaped and supports straplike, olive-green, dichotomously branched fronds. Air bladders develop along the axes of these fronds and the fronds have no midribs. The fronds become fertile in winter, producing lateral receptacles. These succulent winter receptacles, along with the seaweed's abundance across the coastlines of the North Atlantic, mean that it has often been used as fodder or as fertilizer at very local scales. This has seen it recognized with a range of common names: "knobbed wrack," "knotted wrack," "egg wrack," or, in the Orkney islands, "yellow tang" or "paddy tang," "Tang" being the Nordic language root for seaweed.

Pigs, in particular, are said to prefer the taste of *A. nodosum* to that of the kelps, and it has been sold as dried meal in the United States. Various Scandinavian studies in the mid-twentieth century suggested that it was a reasonable way to provide dietary supplementation for livestock, being around three-quarters as effective as hay. It is a good source of alginate and is harvested for alginate extraction. The most promising use of *A. nodosum*, however, is

ABOVE | The fertile reproductive receptacles of *Ascophyllum*.

that it makes an excellent liquid fertilizer. Its extracts provide the same nutrients as those of most other seaweeds, but the thing that sets them apart is that something in the *A. nodosum* extract appears to stimulate the growth of soil fungi, which then help the plant to take up its nutrients. This process remains poorly understood, but much studied.

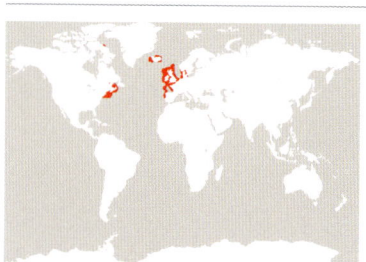

ETYMOLOGY
Greek *askos* = "leathery" + *fyllon* = "leaf"

DIVERSITY
Only one known species

SIZE & SHAPE
Dichotomous; usually an olive-brown color; free-living forms are found on either side of the Atlantic and adopt idiosyncratic morphologies that may not be immediately recognizable as *Ascophyllum*. Some individuals can reach 6½ ft (2 m)

HABITAT
Mid-intertidal zone on sheltered stable shores on either side of the North Atlantic

LIFE CYCLE
Visible diploid adults produce haploid gametes (see page 36)

ABOVE | An underwater canopy of *Ascophyllum nodosum* (Latin: "nodes").

LEFT | The air bladders that form along the axis of *Ascophyllum* blades give it one of its common names: "egg wrack."

PHAEOPHYCEAE: FUCALES: FUCACEAE

FUCUS

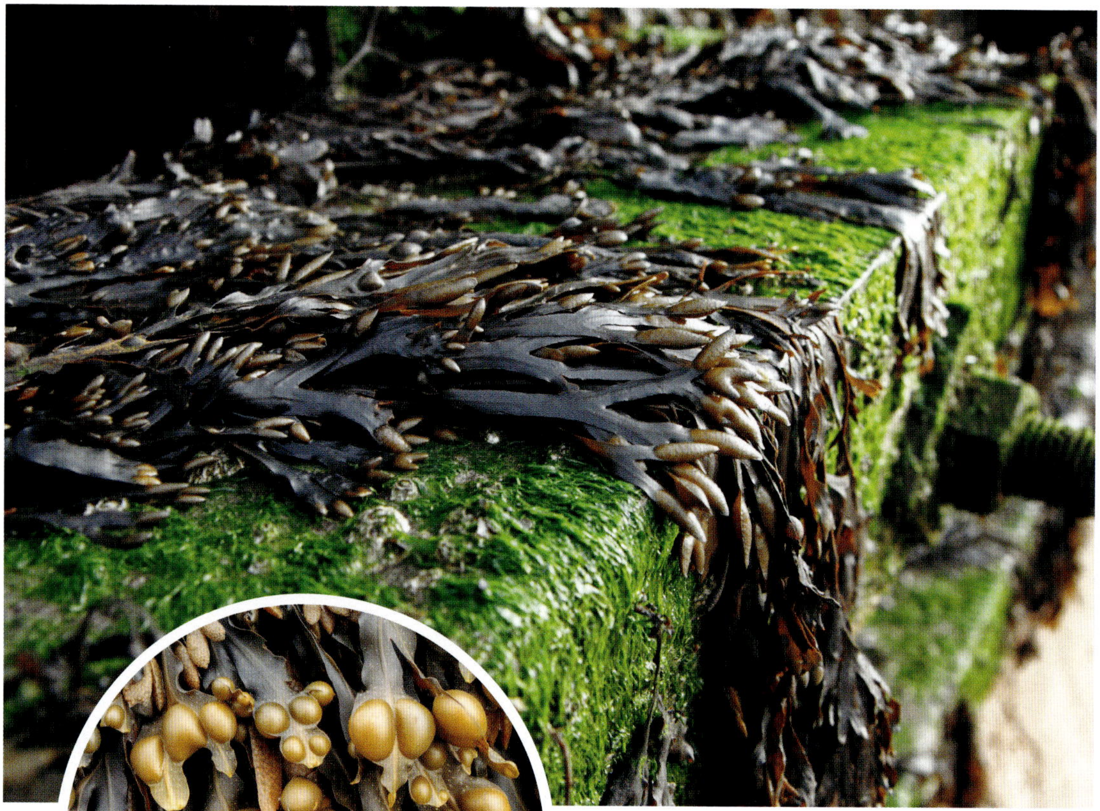

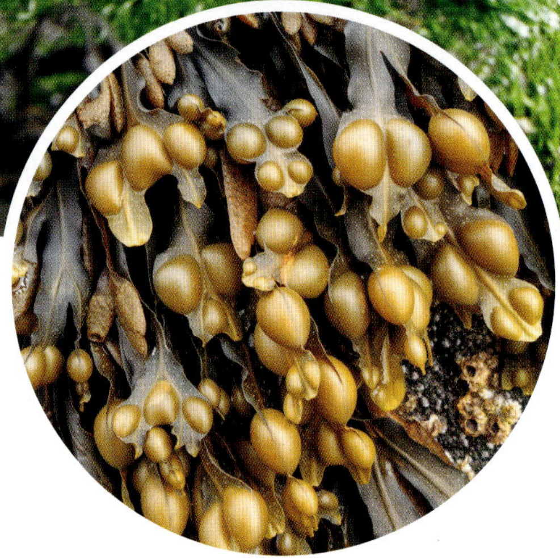

ABOVE | "Spiral wrack." *Fucus spiralis* (Latin: "spiral") growing on concrete at Whitstable on the north coast of Kent, UK. The swollen tips are the fertile receptacles.

LEFT | The distinctive air bladders of bladderwrack, *Fucus vesiculosus* (Latin: "with bladders").

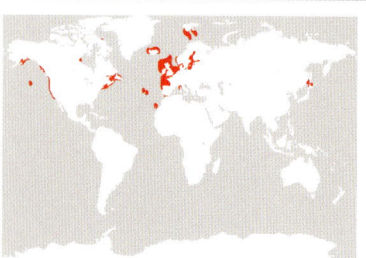

ETYMOLOGY
Greek *phycos* = "seaweed"

DIVERSITY
Contains around 20 accepted species

SIZE & SHAPE
Usually dichotomously branching and olive- to dark-brown in color; some species are only a few inches long, while larger species can reach lengths of 3–6½ ft (1–2 m)

HABITAT
Common above temperate latitudes in the intertidal to near subtidal of the northern hemisphere, with different species growing in overlapping zones; not found in the southern hemisphere.

LIFE CYCLE
The visible fronds are the diploid sporophyte and only the gametes are haploid; some species are monoecious, some are dioecious (see page 36)

The *Fucus* of Linnaeus dominates the sheltered intertidal zones of Atlantic Europe and North America. Commonly known as "wrack," its species are easily gathered and have been used indigenously for hundreds, if not thousands, of years as feed, as fertilizer, and, when dried and burned, as a source of alkaline salts for industrial processes. Several species may be easily distinguished, with a particularly distinctive one being "bladderwrack" (*F. vesiculosus*), which has paired air bladders running up its central midrib. The visible thalli of *Fucus* are the diploid adults and their tough, forked fronds connect through a cordlike stipe to a small holdfast. They grow from apical meristems at the frond tips.

Fucus reproduces sexually and, as with all genera in the order Fucales, the gametes are the only haploid cells in the life cycle. When fertile, gametangia develop in blister-like receptacles at the tips of the fronds. Reproduction is a costly business and, like all seaweeds, *Fucus* species need to maximize the chances of sperm and eggs finding each other on the shoreline, rather than being washed out to sea. To do this, *Fucus* has developed a range of strategies. As with many browns, its eggs release a chemoattractant that guides the sperm toward them. Sperm may contain pigments that allow them to follow light cues and the resulting gametes can be vividly colored, with the bright orange sperm of *F. serratus* being particularly easy to see oozing out of receptacles in rockpools. The timing of gamete release is carefully controlled, with preferential release on calm, sunny days around high tide and with sperm being released before eggs.

ABOVE | "Bladderwrack" or "black tang" or "rockweed": abundant seaweeds like *Fucus vesiculosus* often go by many common names.

PELVETIA

Plants in the *Pelvetia* genus often form the upper zone of brown seaweeds on shorelines around the North Atlantic. The genus has only one major representative, *Pelvetia canaliculata*. This looks very much like smaller thalli in the *Fucus* genus and it took almost a century for *P. canaliculata* to be moved out of Linnaeus' original *Fucus* genus and into its own by Joseph Decaisne and Gustave Thuret in 1845. Its easy availability made *Pelvetia* a common seaweed to harvest for fodder and it was particularly used to fatten pigs along the west coast of Scotland. It is also grazed by coastal ruminants, with the seaweed-eating sheep of the Scottish island of North Ronaldsay being especial gourmets and with the people of the Orkney islands calling it "cow-tang."

As with all the wracks, *P. canaliculata* has a dichotomously branching frond. The fronds are small and compressed, do not have a midrib, and, as the species name suggests, are slightly rolled to form a central channel (hence, *canaliculata*). All of these adaptations lower the effective surface area to volume ratio and make *Pelvetia* species usefully resistant to desiccation. This allows them to live fairly high up in the intertidal zone, where a perennial thallus can usually survive for around five years.

ABOVE | *Pelvetia canaliculata* (Latin: "having channels"), with its fronds rolled around the central channels that give it its name.

Dating back to Thuret's work in the mid-nineteenth century, *Pelvetia* was also a popular model with which to study fertilization and reproduction in the wracks. When fertile, receptacles form at the ends of the fronds. Its gametes may be easily collected, manipulated, and studied under a microscope and both *P. canaliculata* and its *Fucus* cousins have storied histories as models for the study of fertilization and early development.

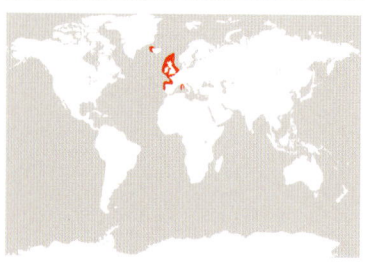

ETYMOLOGY
Named after Dr. Auguste Pelvet, who collected species for Gustave Thuret

DIVERSITY
Only a couple of known species

SIZE & SHAPE
Small and dichotomously branched, usually not being more than 6 in (15 cm) or so

HABITAT
Mid- to high littoral zone across the northern hemisphere, often forming the top zone of brown seaweeds; grows on hard substrates

LIFE CYCLE
Visible diploid adults produce haploid gametes; *P. canaliculata* is monoecious (see page 36)

LEFT | Like most seaweeds, the growth of *Pelvetia canaliculata* is sensitive to environmental conditions. These diploid adults in Scotland have a particularly yellow coloration.

BELOW | Two stands of *Pelvetia* surrounded by the limpets that, given the chance, will graze on any young offspring the seaweed produces.

HORMOSIRA

Plants of the *Hormosira* genus are a common sight along shorelines in the southern hemisphere, to the extent that they are grazed by sheep in New Zealand in much the same way that *Pelvetia canaliculata* is grazed by sheep in Scotland. The plants are perennial and can dominate coastlines. The name *Hormosira* means "necklace chain" and is entirely appropriate, because the dichotomously branching thallus is regularly constricted, dividing what is, in fact, a continuous frond into a chain of bead-sized air bladders. These allow the plants to float and thick carpets of "Neptune's necklace" are a common sight along the coasts of Australia and South Africa. This ability to float also allows fragments of plants to cross large bodies of water and *H. banksii* is thought to have spread outward after originally evolving in Australia. When fertile, most commonly in the southern hemisphere's cooler winter months of July and August, receptacles appear as tiny blisters on the surface of the bladders

As with all brown seaweeds, their growth and development is sensitive to environmental conditions, which means that they can display a great deal of morphological variation. Indeed, they can display so much that it makes it difficult to correctly identify species: a number of *Hormosira* species have been proposed, only for later researchers to collapse them down into morphological variants of the type species *H. banksii*. At the time of writing, *H. pumila* is still accepted as a separate species, but it is possible that this, too, will be found to be a morphotype of *H. banksii* in time.

OPPOSITE, TOP | Thick mats of *Hormosira banksii*, named for Joseph Banks (1743–1820), who was Captain Cook's botanist.

OPPOSITE, CENTER | The air bladders of *Hormosira banksii*, seen here on the northeast coast of New Zealand's South Island, near Kaikoura, Canterbury.

OPPOSITE, BOTTOM | *Hormosira banksii* at Bouddi Point, just north of Syndey, New South Wales, Australia, demonstrating why its common name is "Neptune's necklace."

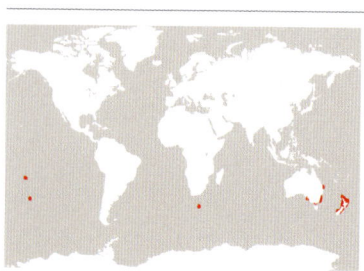

ETYMOLOGY
Greek *hormos* = "necklace" + *seiria* = "chain"

DIVERSITY
Only a couple of known species

SIZE & SHAPE
Distinctive, beads-on-a-chain look; usually light olive to yellow in color and around 12 in (30 cm) long

HABITAT
Southern hemisphere

LIFE CYCLE
Visible diploid adults produce haploid gametes; *Hormosira* is monoecious and can also vegetatively propagate (see page 36)

CHORDA

This genus gives its name to the Chordaceae, one of several important families in the two closely related orders Chordales and Laminariales, which together have traditionally been called the "true kelps." The best-known member of the Chordales is *Chorda filum*, which grows from a disk-shaped holdfast. A beltlike blade (the diploid sporophyte) grows directly from the holdfast as a summer annual, and is lost in the fall. The fronds can grow to several times the length of a person and this impressive size has seen *C. filum* adopted for several uses, under a number of common names: "mermaid's tresses," "bootlace weed," "sea catgut," "lucky Minny's lines," and "dead man's rope." It has been used as a winter feed for sheep and was sometimes used in place of twine around the margins of northwest Europe.

The frond itself is an unbranched hollow cylinder that becomes covered in tiny transparent hairs in summer. Because of their size, the Chordales and Laminariales have the most complex internal architectures of any seaweeds. The blades of juveniles look similar across both orders and develop three layers: a surface meristoderm layer, a cortex underneath the meristoderm, and a central medulla. Cells in the meristoderm are able to divide to repair the frond when the surface is damaged. Cells in the cortex interconnect and some develop into trumpet-shaped connecting cells in the medulla. Finally, cells in the central medulla, including the elongated "trumpet" cells, form internal channels that give kelps a very simple circulatory system. Parts of the frond can become inflated, aiding buoyancy. Under the microscope, this internal structure looks a little like the differentiated parenchyma inside the leaves of land plants (see microscope photograph on page 212), although kelp architecture is not as complex.

OPPOSITE | A canopy of *Chorda filum* (Latin: "thread") in the foreground, off the coast of Dorset, UK, with the invasive wireweed (*Sargassum muticum*) threatening in the background.

OPPOSITE (INSET) | The fronds of the large diploid sporophyte generation of *Chorda filum* are covered in fine hairs.

ETYMOLOGY
Greek *chordi* = "string" or "cord"

DIVERSITY
Five known species

SIZE & SHAPE
Long, unbranched, beltlike fronds that can reach around 33 ft (10 m) long

HABITAT
Cold temperate northern latitudes in both the Atlantic and Pacific, and has penetrated into the Mediterranean and the Baltic seas. Intertidal to subtidal zone, sheltered and sandy or pebbly shores; smaller individuals are sometimes epiphytic

LIFE CYCLE
Heteromorphic alternation of large diploid sporophytes and small haploid gametophytes (see page 37)

UNDARIA

The first major family in the Laminariales is the Alariaceae, named after the genus *Alaria*. Very similar to *Alaria*, *Undaria* is one of the world's most important cultivated seaweeds. The *Undaria pinnatifida* farms of China and Korea harvest several million tons of wet biomass every year— around 5–10 percent of the world's annual seaweed harvest. *Undaria* has been a major food crop around the northwest Pacific for several centuries, where it is eaten as the main species of *wakame*. In Japan, it used to be harvested in summer by fishermen who used corkscrew-like rakes to twist the plants off the seafloor, but most *Undaria* is now farmed. Once collected, it is washed to remove sea salts, dried, and the leafy parts of its fronds are kneaded into wafers that may be eaten as snacks but are most commonly either added to miso soup or turned into the shredded *wakame* that is used as a garnish.

Undaria has long, broad fronds with a wavy edge that gives the genus its name. They are thinner than

ETYMOLOGY
Latin *Unda* = "wave" or "ripple" (as in undulate)

DIVERSITY
Only around four or five species are known

SHAPE & SIZE
A characteristic waved edge to the frond, which is usually less than 10 ft (3 m) long

HABITAT
Subtidal offshore zone in areas of the Pacific that have strong currents

LIFE CYCLE
Heteromorphic alternation of large diploid sporophytes and small haploid gametophytes; dioecious (see page 37)

the fronds of other kelps but are supported by a strengthening midrib. The midrib is jointed rather like a bamboo shoot and is hollow, which helps the fronds to float nearer the surface and the light.

As with other Alariaceae, reproductive sori are produced on specialized fronds called sporophylls, rather than on the main fronds. When *Undaria* becomes fertile, two rows of these sporophylls appear at the base of the fronds and produce the unilocular sporangia that generate the meiospores

that, in turn, grow into the microscopic gametophyte generation. *Undaria* gametophytes can cope with a broader range of temperatures and salinities than those of the other kelps. This means that *Undaria* juveniles can survive when they inadvertently hitch lifts on shipping and this has made *Undaria* an invasive species. It is native to the Pacific, but has now spread to Europe. At the time of writing, it is the only known invasive kelp, although others may follow as climate change redraws our oceans.

OPPOSITE | The base of *Undaria pinnatifida* (Latin: "divided like a feather"), showing the haptera of the holdfast and the elaborate ruffles from which the midrib emerges.

ABOVE | A fertile blade of the closely related *Alaria*, showing the dark patches (sori) that are full of unilocular sporangia.

LEFT | Japanese kelp, *Undaria pinnatifida*, growing off the coast of Shizuoka, Japan.

LAMINARIA

The *Laminaria* genus gives its name to the second major family in the Laminariales, the Laminariaceae, which contains the six genera on the following pages: *Laminaria*, *Macrocystis*, *Saccharina*, *Nereocystis*, *Pelagophycus*, and *Postelsia*.

Kelps in the *Laminaria* genus are usually the largest seaweeds found around most North Atlantic shorelines. Their size and abundance has given them a long history of use and the Pellieux process of 1850 was a precursor of today's biorefinery approaches. *Laminaria* was chopped and ensiled, with iodine coming out during the dewatering. The remaining mass was fermented, treated with lime,

and burned. The residual ash was then dissolved to give potash and soda, so that multiple products could be extracted from the one kelp. Much effort is currently going into seeing whether a similarly efficient process can be used to collect alginate and mannitol, while converting the rest of the *Laminaria* biomass into biofuel.

More recondarily, in the early twentieth century, short fragments of *Laminaria* stipe were used to widen small holes during surgery, because the material of the stipe swells in a controllable fashion when water is added. Similar ideas are being used today in the field of biomimicry, in which problems

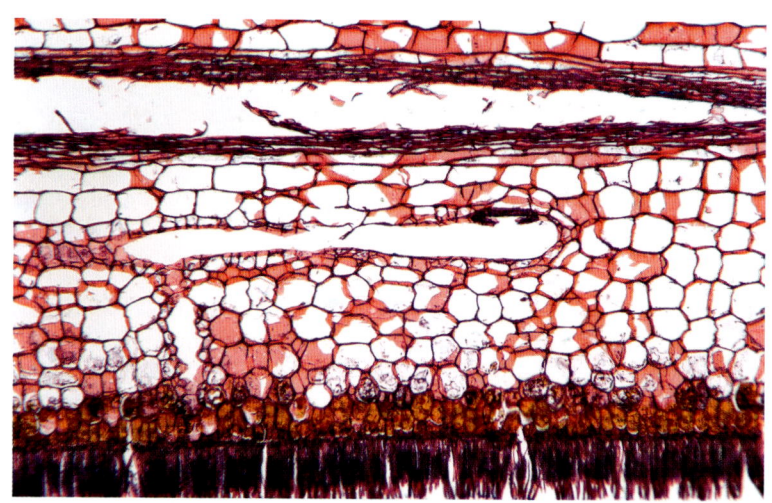

LEFT | A cross section through a *Laminaria* blade, showing the outer cortical layer (bottom) and the more loosely packed medulla (middle).

OPPOSITE | The tough stipes of *Laminaria hyperborea* (Latin: "northern") standing upright at low tide and trailing their blades.

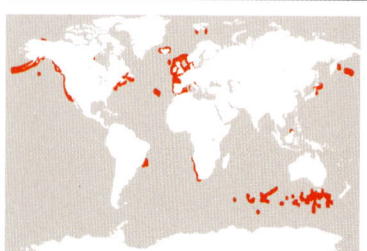

ETYMOLOGY
Latin *lamina* = "blade"

DIVERSITY
Around 40 known species, with perhaps five to ten more to be delineated

SHAPE & SIZE
A broad, wide blade, usually several feet long and divided into fingers

HABITAT
Rocky subtidal zone, offshore, and out to around 65 ft (20 m) depth in cold and temperate northern hemisphere waters

LIFE CYCLE
Heteromorphic alternation of large diploid sporophytes and small haploid gametophytes (see page 37)

are solved by looking at the structures of natural products with the desired properties.

Thalli in *Laminaria* are all perennial, although wave action will often scour the old, deep brown fronds away and newer, lighter ones will regenerate from the holdfasts to take their place. The fronds lack midribs and are usually divided into broad, distinctive, straplike fingers. However, as with all seaweeds, their growth is sensitive to environmental conditions, so *Laminaria* species can display a lot of morphological variation. Different species can still be distinguished if you know how: the stipe of *L. digitata*, for example, has an oval cross section while the stipe of *L. hyperborea* has a circular cross section. The holdfasts in *Laminaria* are made of tangles of tubular bodies called haptera. The dried and whitened holdfasts, usually with a short stump of stipe still protruding, are a common sight on Atlantic beaches.

The *Laminaria* genus has given its name to both its family and its order and all members of that order grow in the same way. The main intercalary meristem lies in the transition zone, where the stipe changes into the frond, and growth from this main meristem drives the elongation of the blade. However, as explained in the entry for *Chorda* (see page 208), members of the Laminariales have a secondary surface meristem, called the meristoderm, which acts to repair the damaged surface of fronds.

As with all the kelps, *Laminaria* is the focus of much environmental worry because its species have evolved in cold or temperate waters and climate change is now rapidly warming those waters. F orests of *Laminaria* have supported Atlantic coasts for thousands of years, so the range shifts and reorganizations of *Laminaria* that we are now seeing are a real and increasing concern.

MACROCYSTIS

BELOW | *Macrocystis pyrifera* (Latin: "pear carrier," because of the air bladders) off Santa Barbara island, California, USA.

In his 1839 *Voyage of the Beagle*, Charles Darwin wrote: "There is one marine production which, from its importance, is worthy of a particular history. It is the kelp, or *Macrocystis pyrifera* [. . .] The number of living creatures of all orders, whose existence intimately depends on the kelp, is wonderful." It is hard to disagree. This truly

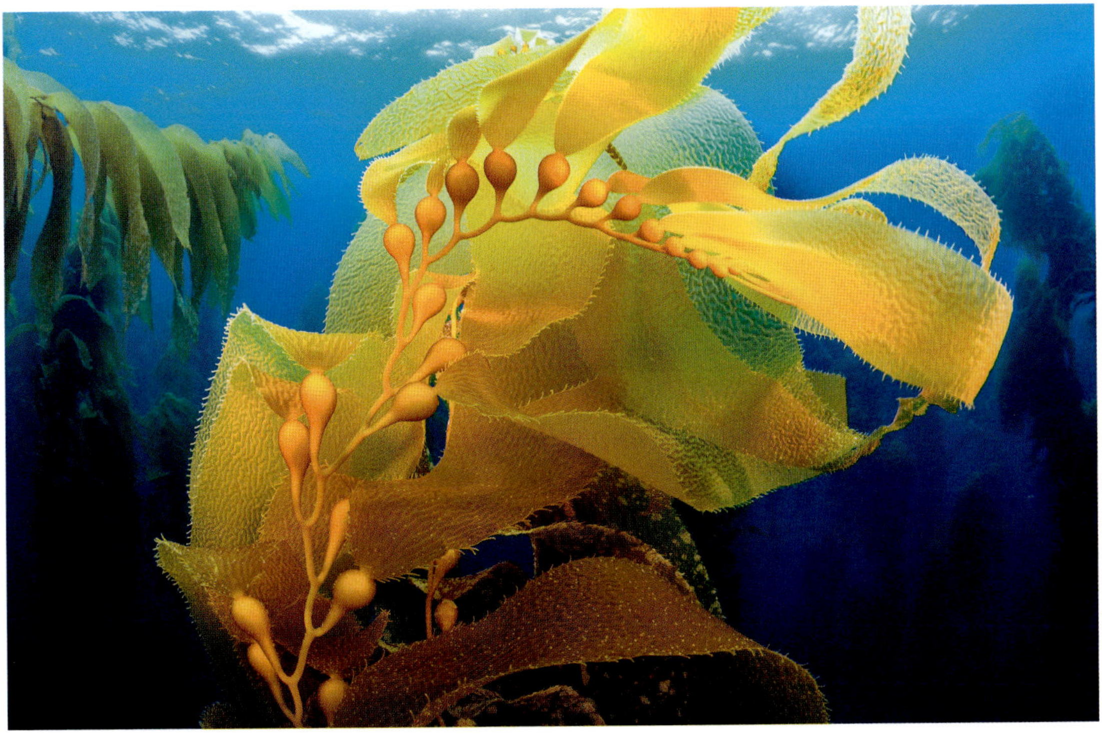

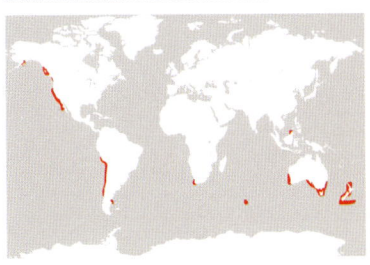

ETYMOLOGY
Greek *makros* = "long" + *kystis* = "pouch"

DIVERSITY
One known species

SHAPE & SIZE
The longest known seaweed species; fronds can be longer than 165 ft (50 m) and up to around 90 lb (40 kg) in wet weight

HABITAT
Offshore subtidal zone in cold temperate waters of the north and south hemispheres

LIFE CYCLE
Heteromorphic alternation of large diploid sporophytes and small haploid gametophytes (see page 37)

spectacular genus was first described by Europeans in the sixteenth century, although it was obviously known and used earlier by indigenous peoples. The Maori, for example, ground it into a meal.

The fronds of the giant kelp are annual and last for around six months. They are rugose (wrinkled) and dentate (having tiny, toothlike protrusions on their edges), although there is much morphological variation. This complicates their study: around a dozen different species of *Macrocystis* have been proposed but they all seem to be morphological variants of *M. pyrifera*, which, at the time of writing, remains the only confirmed species in this genus. The holdfast can be several feet wide. The stipe emerges from the holdfast, divides, and each of the two resulting branches develops a blade, which is called the frond initial. The stipe then elongates and new lateral branches arise from the meristem at the base of the frond initial. Each new branch develops a basal air bladder. The overall effect is of a curved stipe with multiple blades rising from it at regular intervals. These blades are annual and grow extremely fast, with some growing at almost 20 in (50 cm) per day to reach lengths upward of 165 ft (50 m). When fertile, paired sporophylls develop near the frond initials.

Its large size and rapid growth made *Macrocystis* one of the staples of the US West Coast potash boom of the early twentieth century and the genus is now being looked at as a potential food and biofuel crop. In contrast to the blades, the base of the giant kelp is perennial and the plants live for several years,

so older plants can support many tangled stipes. This ability of the base to rapidly regenerate new stipes and blades has led to speculation that *Macrocystis* could be harvested twice a year.

More importantly, and as Darwin noted, giant kelp forests support a huge range of marine life. A salutary example are the sea otters that swim among the forests off the California coast and feed on sea urchins. Otter populations crashed in the 1850s because they were hunted to support the fur trade. The giant kelps then became overgrazed by sea urchins. The recovery of the sea otter population has helped to stabilize this remarkable and important ecosystem, but climate change looms as a new and potent threat.

PHAEOPHYCEAE: LAMINARIALES: LAMINARIACEAE

SACCHARINA

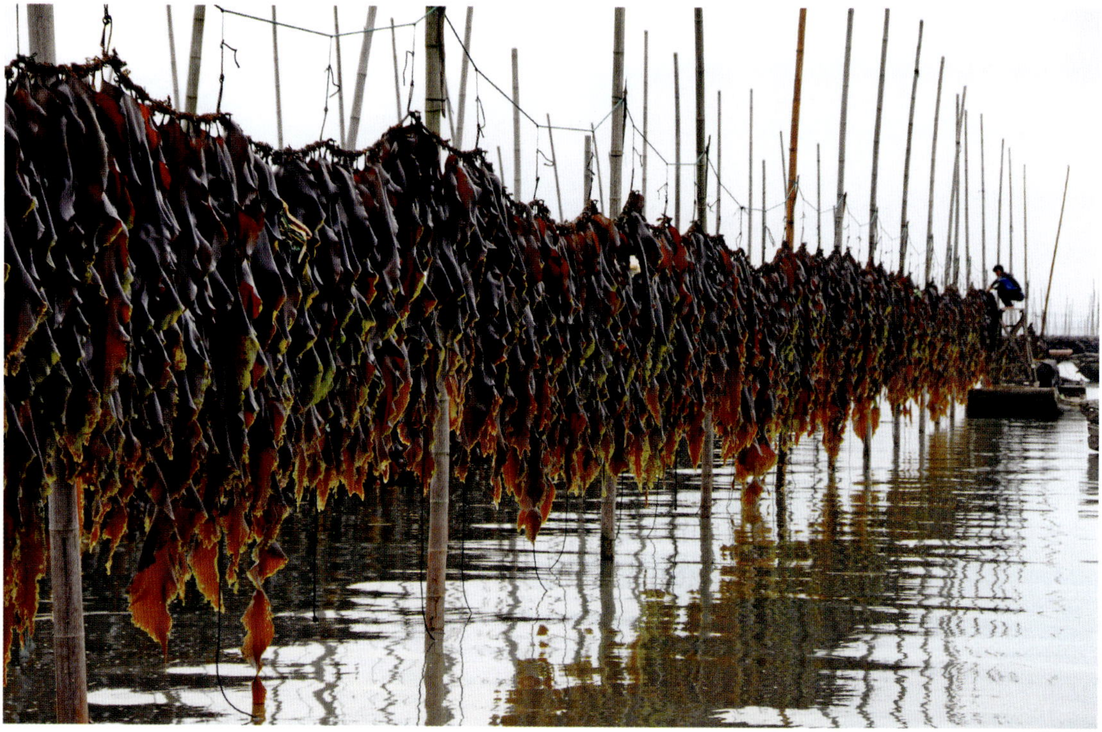

This third genus in the Laminariaceae is *Saccharina*. *Saccharina* fronds are annuals that become fertile in summer. The fronds form long, single blades, often with ruffled edges and a central corrugated band that runs the length of the frond. Their growth is environmentally sensitive, however, so the genus shows significant growth variation across its range. The ruffles can depend on temperature, for example, with smoother fronds seen in warmer temperatures.

The genus' most notable member is undoubtedly *S. japonica*, which is currently the world's most harvested seaweed. *S. japonica* has unusually high levels of the photosynthetic product, mannitol.

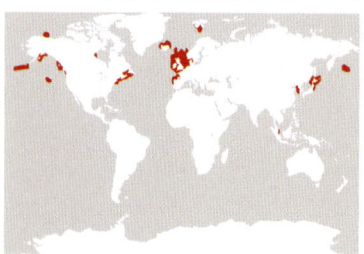

ETYMOLOGY
Modern Latin *saccharum* (probably a loan word through multiple languages from an original Sanskrit word) = "sugar"

DIVERSITY
Around 20 species are known but a number of species were reassigned from the *Laminaria* genus into the *Saccharina* genus in the early twenty-first century, so many references to *Laminaria japonica* (for example) may still be found

SHAPE & SIZE
Usually less than 65 ft (20 m) long

HABITAT
Stones or shingle, can be found 80 ft (25 m) deep

LIFE CYCLE
Heteromorphic alternation of large diploid sporophytes and small haploid gametophytes (see page 37)

ABOVE | A plate of prepared and shredded *Saccharina*, known as *haidai* in China, *dasima* in Korea, and *kombu* in Japan.

This can taste umami and a little like peanuts, but is also sweet, which explains both the name of the genus (and the common name of "sugar kelp") and the long history of eating *Saccharina* in the countries around the northwest Pacific. It has been the *kombu* of Japan since at least the eighteenth century, the *haidai* of China, and the *dasima* of Korea.

Historically, *S. japonica* was harvested especially off the northern Japanese island of Hokkaido where the cold Arctic current favors their growth. However, the cultivation of kelp gametophytes has been known since about the mid-twentieth century and the pioneering Chinese botanist, Cheng Kui Tseng, developed that knowledge into the Chinese maricultural revolution of the 1950s and 1960s. Tseng decided to focus on the cultivation of *Undaria* and *S. japonica* and more than 10 million tons of *Saccharina* are now harvested every year, almost all of it in China, North Korea, South Korea, and Japan. The dried fronds of *S. japonica* are a versatile ingredient: they go into broths and flavor kelp tea (the *kombu-cha* of Japan). They may also be pickled, deep-fried, or just eaten intact, usually sweetened and slightly caramelized.

NEREOCYSTIS

When Elizabeth Smart, in her poetic 1945 novel *By Grand Central Station I Sat Down and Wept*, wrote that: "The kelp in amorous coils appear to pin down the Pacific," she was probably writing about *Nereocystis*. The single species in the genus, *N. luetkeana*, was the most common kelp off Smart's British Columbia when she was writing in the 1940s and the species is one of a couple to have the common name "bull kelp."

Nereocystis is one of the kelp genera in which the fronds divide through the transition zone to create multiple meristems. The holdfast is a hemispheric mass of haptera that spans around 12 in (30 cm) in diameter. A watermelon-sized air bladder grows at the end of the stipe and this bladder is filled with gases, especially carbon monoxide, which aids flotation. Four short, flattened extensions develop from this bladder and each extension branches dichotomously into blades. A single *Nereocystis* plant can end up with around 100 blades, although many have far fewer. The blades are annuals. When the plant becomes fertile, the blades turn into sori and their color leaches away as the meiospores swim off, leaving the now-dead, whitish, cell-wall scaffold to slowly disintegrate.

Its abundance and size made *Nereocystis* one of the main Pacific potash species in the early twentieth century and its high levels of alginate make the genus an attractive possible source of biopolymers today. There have been some attempts at broader commercial use. In the early twentieth century, for example, researchers at the University of Washington patented a candied *Nereocystis* snack called "Seatron," which never really took off in the same way as the candied seaweed snacks of Japan and Scotland.

The indigenous peoples of Alaska have used this seaweed to better effect, adapting the stipe of *Nereocystis* into a fishing line. They also used the hollow stipes as tubing in the stills in which they distilled the alcoholic liquor called hoochinoo, whence the common US slang "hootch" for dubiously strong alcohol. Perhaps more impressively, indigenous peoples are reported to have heated the *Nereocystis* bladder while putting the stipe in their ear. The steam coming out was believed to cure headaches.

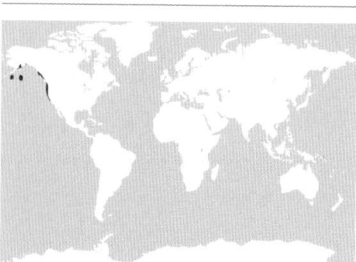

ETYMOLOGY
From Nereus, a Greek god of the sea + Greek *kystis* = "pouch"

DIVERSITY
Only one known species

SHAPE & SIZE
Larger individuals have been measured at around 131 ft (40 m) and with wet weights of 55 lb (25 kg)

HABITAT
Colder waters of the northeast Pacific; deep, out to around 100 ft (30 m) depth

LIFE CYCLE
Heteromorphic alternation of large diploid sporophytes and small haploid gametophytes (see page 37)

ABOVE | The other "bull kelp" (see *Durvillaea* on page 195). Blades of *Nereocystis luetkeana*, named for the German Russian Arctic explorer Friedrich von Lütke (1797–1882). These are floating off Vancouver Island, British Columbia, Canada.

RIGHT | An *Nereocystis* air bladder being held upside down. The stipe is at the top and the four main blades can be seen at the bottom (some are branching).

PELAGOPHYCUS

Some of the earliest records of the Pacific genus, *Pelagophycus*, may be squiggles on sixteenth-century Spanish maps. *Pelagophycus*, and the other kelps, were used as a sign that land was nearby and, as importantly, to show where offshore rocks lay. *Pelagophycus* usually grows in deep water, but populations will grow in much shallower waters during el Niño years. *Pelagophycus* is thought to be a monotypic genus, which means that it only contains one species: *P. porra*. However, one extremely interesting thing about this kelp is that windward and leeward populations exist. Windward plants tend to be larger and perennial, while leeward plants tend to be smaller annuals. These patterns suggest that *P. porra* is slowly beginning to split into two species, offering an exciting window into speciation.

Morphologically, *Pelagophycus* is reasonably similar to *Nereocystis*. In both, the holdfast is made of a hemispherical mass of haptera and the long stipe ends in a watermelon-sized air bladder from which a number of blades arise. When fertile, sori form across the blades. In *Nereocystis* the stipe is hollow and there are four outgrowths from the bladder that have very short connections before the fronds begin, so that the blades look like they spring directly from the bladder. In contrast, in *Pelagophycus* the stipe is solid enough to be carved as driftwood curios and there are only two much longer outgrowths, so that two clear branches may be seen. Each of the two branches grows in a sympodial fashion, which means that each branch grows one new branch, which in turn grows one new branch, and so on. Each new branch terminates in a corrugated blade that has small spines around its edge. The fronds can be very large, often reaching 30 ft (10 m) in length. The resulting structure gives an arc-shaped main axis with regular branches on only one side of the axis. Or, to write that more prosaically, the two branches look like the antlers of an elk, which is where this species gets its common name: "elk kelp." The species has never been of major commercial importance, although it was, like other kelps of the Californian coastline, occasionally pressed into service in potash manufacture.

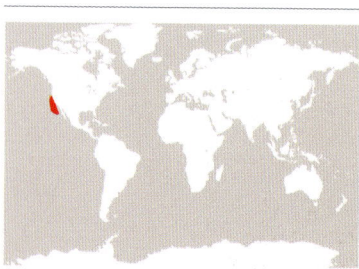

ETYMOLOGY
Greek *pelagos* = "float" + *phycos* = "seaweed"

DIVERSITY
One known species

SHAPE & SIZE
A distinctive, "elk's head" shape, with fronds reaching 131 ft (40 m) in length and a wet weight of 55 lb (25 kg)

HABITAT
Deep subtidal waters off Baja California, around 165 ft (50 m) deep; it grows in more shallow waters during el Niño years

LIFE CYCLE
Heteromorphic alternation of large diploid sporophytes and small haploid gametophytes (see page 37)

ABOVE | "Elk kelp." *Pelagophycus porra* off Santa Barbara Island, California, USA. The sympodial growth of each branch is visible.

POSTELSIA

The one known species in the *Postelsia* genus is, quite possibly, the toughest seaweed out there. It is only found in heavy surf along the Pacific coast but is extremely distinctive.

Postelsia starts, as do all the kelps, from a leaflike juvenile. A strong hemisphere of haptera develops and fastens the thallus tightly to the rock on which it sits. The stipe of *Postelsia* thickens into a tough vertical tube. The fronds are annuals and divide down through the transition zone to give multiple meristems, often resulting in 100 or more blades that hang down from the erect stipe. The blades have

sawtooth edges and are corrugated. The overall effect is that of a small palm tree, which gives the one species in the genus its name: *Postelsia palmaeformis* (Latin = "palm form"), or "sea palm."

This unusual shape is matched by an unusual lifestyle. *Postelsia* is, for want of a better term, a glutton for punishment. Its hardy stipe and tough holdfast allow it to withstand extreme wave scouring that removes other seaweeds and mussels. It can survive out of water for much longer than other kelps. *Postelsia* is therefore able to settle and colonize rocky shores that would be too inhospitable for most other organisms. Indeed, it acts as the host for other epiphytic seaweeds such as filamentous *Pylaiella* species (see page 182) that would otherwise be unable to cling to the rocks. When fertile, sori form in the grooves of the blades and the resulting gametophytes tend to settle quickly and then release their gametes at low tide to avoid having them washed out to sea by the rough surf. As a result, *Postelsia* plants tend to be found in clumps.

LEFT | *Postelsia palmaeformis* (Latin: "palm-shaped") growing on exposed rocks on the coast of Vancouver Island, British Columbia, Canada.

OPPOSITE | Young diploid sporophytes of *Postelsia palmaeformis* at Cape Perpetua, Oregon, USA, their hapteroid holdfasts clearly visible.

ETYMOLOGY
Named after Alexander Postels (1801–71), a Russian naturalist of Baltic German extraction who was a colleague of the person who described the genus

DIVERSITY
One known species

SHAPE & SIZE
Can be 40 in (100 cm) high, although most are more around 12 in (30 cm)

HABITAT
Mid- to lower intertidal in heavy surf along the temperate Pacific coast

LIFE CYCLE
Heteromorphic alternation of large diploid sporophytes and small haploid gametophytes (see page 37)

EISENIA

The third major family in the Laminariales is the Lessoniaceae. Originating in the Pacific they include the three genera on the following pages: *Eisenia*, *Ecklonia*, and *Egregia*.

Eisenia was first described when the Swedish-American Gustav Eisen (1847–1940) collected an unusual looking kelp from Santa Catalina Island, south of Los Angeles. The kelp was growing on rocks in strong surf and its stipe was stiff enough to hold its fronds up near the light when submerged, but also flexible enough to bend significantly in waves. Eisen sent the specimen to a fellow Swede at the University of Uppsala, Johan Erhard Areschoug (1811–87), who promptly named the genus after Eisen.

The genus contains "sea oaks" (*E. bicyclis*), Southern "sea palms" (*E. arborea*), and the *arame* of Japan, cultivated in both Japan and Korea and

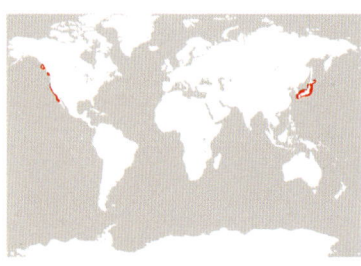

ETYMOLOGY
Named after Gustav Eisen (1847–1940)

DIVERSITY
Only around five to ten species are currently recognized

SHAPE & SIZE
Usually less than 40 in (100 cm) in length

HABITAT
Low to subtidal zone along the eastern Pacific coast

LIFE CYCLE
Heteromorphic alternation of large diploid sporophytes and small haploid gametophytes (see page 37)

used in soups or as a jelly in soybean sauce. It is also a good source of alginate and one of the species used in its extraction. More oddly, it is sometimes hung around doors and windows as a Japanese New Year decoration because the fronds turn a pleasant shade of green as they dry.

Eisenia has an unusual growth pattern. Two blades develop from the stipe and are eaten away by erosion or grazing. From each of their flattened stubs, multiple reproductive sporophylls develop and a single plant can have 30–50 sawtoothed sporophyll blades. Sori develop on these sporophyll blades and are made of microscopic unilocular sporangia that grow out from the surface of the fronds. The unilocs are protected by a palisade of specialized cells that grow next to them and are called paraphyses. Common to all Laminariales, paraphyses are not found in any other seaweeds.

OPPOSITE | The diploid sporophytes of *Eisenia arborea* (Latin: "tree"), clearly showing the forked stipe with a blade on each fork. This stand of *Eisenia* is being threatened by the sea urchins below, which feed on them.

LEFT | *Eisenia bicyclis* (Latin: "two wheels," in reference to the platelike sporophylls) growing in a tidepool on the south coast of Japan.

K elps of the *Ecklonia* genus, especially *E. cava* and *E. radiata*, make up the major part of kelp beds in Australia, New Zealand, and South Africa, where it is sometimes called "sea bamboo."

The plants are perennials and are attractively constructed. They begin as a very standard hapteroid holdfast with a hollow, gas-filled stipe and a long primary blade, but lateral secondary blades grow out from the side of the primary blade to give a fairly wild-looking profusion of broad, sawtoothed blades. An air bladder is found at the base of the primary blade, and the bladder and the hollow stipe both mean that *Ecklonia* plants will often float up to form the canopy of their kelp beds.

Ecklonia is commonly eaten in Japan, where it is one of the *kombu* species (*kajime*), and China. It is rich

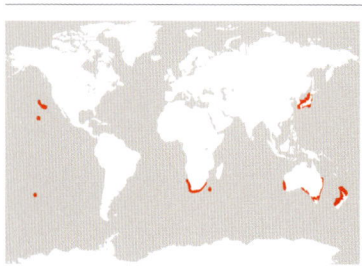

ETYMOLOGY
Named after Christian Friedrich Ecklon (1795–1868), a Danish botanist who was particularly active in collecting specimens from South Africa

DIVERSITY
Around ten known species

SHAPE & SIZE
Larger species (for example, *E. maxima*) can be 50 ft (15 m) long, although most species are 3–6½ ft (1–2 m)

RANGE
Shallow subtidal zones across the Pacific and Indian oceans, down to New Zealand and South Africa

LIFE CYCLE
Heteromorphic alternation of large diploid sporophytes and small haploid gametophytes (see page 37)

in iodine, especially in summer when it is usually harvested, and was one of the staples of the Japanese iodine production industry in the early twentieth century. It has long been used in traditional Korean and Chinese medicine to treat various conditions and is known to produce a range of antioxidant chemicals, which the seaweed presumably makes to discourage sea life from grazing on it. At the time of writing, it is seeing increasing use as a supplement worldwide, although many things produce antioxidants and the ones from *Ecklonia* are unlikely to be much better or worse than those from other sources. It is also an occasional source of alginate.

OPPOSITE, LEFT | *Ecklonia radiata* growing on the side of a jetty in Rapid Bay, just south of Adelaide, South Australia, and showing their sawtoothed blades.

OPPOSITE, RIGHT | Blades of *Ecklonia maxima* (Latin: "big") growing off Buffelsbaai, Western Cape, South Africa, with mats of the brown fucoid seaweed *Bifurcariopsis capensis* underneath.

ABOVE | "Sea bamboo." A canopy and undercanopy of stipes and blades of *Ecklonia radiata*, showing how these kelps can form offshore forests.

EGREGIA

As the genus name suggests, *Egregia* species look egregiously different to most other kelps. Its common name of "feather-boa kelp" is quite accurate, because the overall impression is of an underwater feather boa. To create this unusual thallus, the initial juvenile develops from a conical hapteroid holdfast. The stipe then develops into a flattened, straplike backbone from which very many smaller and opposed stalks arise at regular intervals. This branching pattern is unusual for kelps, which normally have dichotomous or simple branching patterns. The dark brown frond is relatively narrow, often corrugated, and ends in a flattened terminal blade.

Each of the small stalks on the frond develops into an organ. Some become air bladders, some—when the plant is fertile—become its unilocular sporangia, while others remain as stalks that may be narrow or flattened into lobes.

Egregia is perennial and is conspicuous along much of the length of the Pacific coast of North America. It displays a lot of morphological variation, so that plants in warmer waters can look quite different to those in cooler waters. This has made species assignation difficult and, at the time of writing, only two species are known in this genus, with the more common being *E. menziesii*.

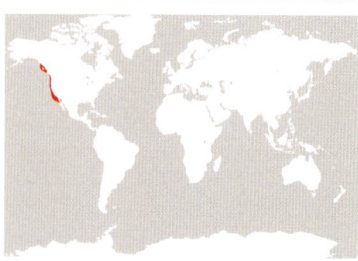

ETYMOLOGY
Latin *egregius* = "remarkable"

DIVERSITY
Only two known species

SHAPE & SIZE
Like a large and leathery feather boa; larger individuals can reach 25 ft (8 m)

HABITAT
Lower to subtidal zone across the western Pacific; can be exposed

LIFE CYCLE
Heteromorphic alternation of large diploid sporophytes and small haploid gametophytes (see page 37)

ABOVE | The diploid sporophytes of *Egregia menziesii*, named for Archibald Menzies (1754–1842), who was Captain Vancouver's naturalist during HMS *Discovery*'s trip around the world in the late 1700s. Appropriately, these are growing off the west coast of Vancouver island, British Columbia, Canada.

RIGHT | The frond of *Egregia menziesii*, showing the many opposed branches that can turn into air bladders or reproductive organs.

GLOSSARY

Agar (after its name in Malay, which is *agar agar*): a red seaweed product that is used as a thickener in the food and drug industries. It is similar to **carrageenan**.

Anisogamous (Gr. "gametes that are not the same size"): having slightly different sized **gametes**.

Antheridium (Gr. "little male"): the structure that makes male sperm in **species** that have different sized reproductive cells (i.e., **oögamous** species). It is one type of male **gametangium**.

Apex, apical (L. "tip"): at the end of something. If a seaweed grows only at the tips of its fronds, we say it shows apical growth.

Archaea (Gr. "ancient"): the lesser-known cousins of bacteria. Like bacteria, they are single celled and have no nucleus.

Axis (L. "axle"): a line that something revolves around, such as Earth's axis. The central "trunk" of a seaweed is the primary axis and branches are secondary axes.

Billion (French "the second step after a million"): A thousand million, so 1,000,000,000.

Blade (Old English): any flat part of a seaweed.

Carpogonium (Gr. "fruit that gives offspring"): a red seaweed structure that makes female eggs. It is the **oögonium** in red seaweeds.

Carpospore (Gr. "seed from the fruit"): the asexual reproductive cells produced in the **carposporophyte** by cell replication (**mitosis**). It develops into the **tetrasporophyte**.

Carposporophyte (Gr. "fruit-bearing plant"): a **diploid** multicellular life cycle stage in the red Florideophyceae that develops from the fusion of male **spermatia** and female eggs.

Carrageenan (after its name in Irish Gaelic): a red seaweed product that is used as a thickener in the food and drug industries. It is similar to **agar**.

Chloroplast (Gr. "colored thing"): the part of a cell that carries out **photosynthesis**. Chloroplasts are found in eukaryotic algae and land plants and are derived from **cyanobacteria** by **endosymbiosis**.

Class (L. "a division"): a large taxonomic group that sits below a **phylum** and above an **order**. See page 74.

Conceptacle (L. "holds things together"): a tiny pit that develops in fertile seaweeds and contains several **gametangia**. Not to be confused with a **receptacle**.

Conchocelis phase (Gr. "shell worm"): a **diploid** multicellular life cycle stage in the red order Bangiales that develops from the **zygotospores** released by the fertilized **carpogonium** and which, in turn, produces **conchospores**.

Conchosporangium (Gr. "seed-bearing thing on a shell"): a structure that develops from the fertile **conchocelis** to make **conchospores**.

Conchospore (Gr. "seed from the shell"): the asexual reproductive cells that are released by the **conchosporangium**. Once released, they undergo cell division (**meiosis**) to become haploid and then grow into the **gametophyte** generation.

Cortex, cortical (L. "bark of a tree"): the outer cell layers of a seaweed blade. The inner layers make up the **medulla**.

Crustose (Middle English): looking like a crust.

Cyanobacteria (Gr. "blue-green little rods"): a group of bacteria that evolved photosynthesis. They are often called the "blue-green algae" but are bacterial and not eukaryotic.

Dichotomous (Gr. "cutting in two"): a symmetrical division into two equal parts. A capital letter Y is dichotomously branched.

Dioecious (Gr. "having two houses"): having separate male and female individuals at some point in the life cycle.

Diploid (Gr. "double"): having two complete sets of all the genes needed to describe an organism. One set comes from each parent. Humans are diploid.

Distromatic (Gr. "two layers"): a seaweed **thallus** made from two layers of cells, rather like a sandwich. See page 78.

Endosymbiosis (Gr. "living inside and together"): one cell living inside another cell, where both cells benefit from living together (as opposed to parasitism, where only one cell benefits).

Epiphyte (Gr. "an additional plant"): a seaweed or plant that grows on another organism but is free-living and not parasitic.

Family (Middle English "a household"): a good-sized taxonomic group that sits below an **order** and above a **genus**. See page 74.

Flagellum (L. "a whip"): a microscopic "tail" on a cell that can help to move the cell around.

Frond (L. "leaf"): loosely used to mean the body of a seaweed.

Gametangium (Gr. "a container"): a structure that makes sexual reproductive cells called **gametes**.

Gamete (Gr. "spouse"): the sexual reproductive cell that usually (but see **parthenogenesis**) needs to combine with another reproductive cell to form a **zygote** before growing into a multicellular seaweed (as opposed to a **spore**). In humans, sperm are male gametes and eggs are female gametes.

Gametophyte (Gr. "the plant that makes gametes"): the life cycle generation that produces **gametes**, usually in its **gametangia**.

Genera (Gr. "types"): the plural of **genus**.

Genus (Gr. "a type"): a relatively small taxonomic group that sits below a **family** and above a **species**. See page 74.

GYA: An abbreviation of Giga (= one **billion**) Years Ago, used to mean either one billion years ago or one billion years old.

Haploid (Gr. "single"): having only one complete set of all the genes needed to describe an organism. Haploid cells are made from **diploid** ones by **meiosis**.

Haptera (Gr. "a fastening"): fingerlike tubes that form the **holdfasts** of some brown seaweeds to give a tangled, fistlike holdfast (see page 79).

Heteromorphic (Gr. "different shapes"): having two different shapes. It is usually used to refer to seaweeds that have different growth forms in their **gametophyte** and **sporophyte** generations.

Heterotrichy (Gr. "different hairs"): a type of growth in which a single-celled **zygote** divides into two filamentous cells. One filamentous cell attaches flat to the thing the seaweed is growing on and the other filamentous cell grows erect and eventually turns into fronds (see page 79).

Holdfast (Old English): the part of the seaweed that attaches it to something.

Holocarpic (Gr. "whole fruit"): a kind of reproduction in which an entire fertile seaweed turns into reproductive cells, rather than just having some bits of it produce them.

Hydrocolloid (Gr. "water glue"): a chemical that absorbs water to become a gel. They are usually **polymers**.

Intercalary (L. "into the calendar"): originally meant a day that was inserted into the calendar to bring it into line with the solar year. A leap day is an intercalary day, for example. In seaweeds, it means growth from the middle of the seaweed, and is usually contrasted with **apical** growth (see page 79).

Isogamous (Gr. "gametes that are the same size"): having **gametes** that are the same size.

Isomorphic (Gr. "same shape"): two things that are the same shape. It is usually used to refer to seaweeds that look the same in both their **gametophyte** and **sporophyte** generations.

Kingdom (Old English "a territory"): the largest taxonomic group, sitting above a **phylum**. See p. 74.

Mastigoneme (Gr. "threads on a whip"): microscopic hairs that grow from some **flagella**.

Medulla (L. "marrow"). The inner, central part of a seaweed blade. The outer part is the **cortex**.

Meiosis (Gr. "a lessening of something"): cell division in which a **diploid** cell is divided into two **haploid** cells.

Meiospore (Gr. "seed produced from a lessening"): reproductive cell formed by cell division (**meiosis**).

Meristem (Gr. "divisible"): a cluster of cells that are actively generating new tissue through cell replication (**mitosis**).

Meristoderm (Gr. "divisible skin"). A layer of cells on the surface of a seaweed that are able to generate new cells by cell replication. Not all **species** have them.

Mitochondrion (Gr. "thread-like grain"): the part of a cell that carries out **respiration**. They are found in almost all eukaryotes and are derived from (α-proteo) bacteria by **endosymbiosis**.

Mitosis (Gr. "having threads"): cell replication. A **haploid** cell replicates to give two haploid daughters; a **diploid** cell will replicate to give two diploid daughters. The name comes from microscopic DNA threads, called chromosomes, that appear in the cell nucleus during replication.

Mitospore (Gr. "seed produced from having threads"). A reproductive cell formed by cell replication (**mitosis**).

Monoecious (Gr. "having one house"): having both male and female reproductive structures on the same seaweed.

Monopodial growth (Gr. "single-footed"): a type of growth in which the main branch (the primary **axis**) of the seaweed grows in one direction and lateral branches (or secondary axes) grow off the main branch.

Monosiphonous (Gr. "a single pipe"): a seaweed body plan in which each **axis** is only one cell thick.

Monostromatic (Gr. "one layer"): a seaweed **thallus** made from one layer of cells. See page 78.

Multiaxial (L. "many directions"): a seaweed with a body plan made up of lots of filaments that all start from the same place. This is sometimes called "fountain-like" growth (see page 78).

Oögamous (Gr. "gametes that are eggs"): having **gametes** that are either small sperm or large eggs.

Oögonium (Gr. "egg maker"): the structure that makes eggs in **species** that have different sized reproductive cells (i.e., **oögamous** species). It is one type of female **gametangium**.

Order (L. "a rank"): a large taxonomic group that sits below a **class** and above a **family**. See page 74.

Paraphysis (Gr. "the thing next to the thing giving birth"): a cell that grows next to **sporangia** in **sori** and which helps to support and protect the sporangia themselves.

Parenchyma (Gr. "the stuff that is poured in"): a type of internal architecture seen in land plants in which the plant is made of neatly packed, regularly shaped cells that have different and distinct functions. Seaweeds do not have this, although the kelps almost do.

Parthenogenesis (Gr. "virgin birth"): when a **gamete** grows into a seaweed without having been fertilized.

Photoautotrophy (Gr. "feeding yourself with food made from light"): the process of making sugars from **photosynthesis** and then powering yourself by burning those sugars in **respiration**.

Photosynthesis (Gr. "making things from light"): the biological process by which algae and plants make sugars from carbon dioxide and water, using light to power the reaction.

Phylogeny (Gr. "the making of a tribe"): the evolutionary "family tree" of a **species**, sometimes called a phylogenetic tree.

Phylum (Gr. "a tribe"): a very large taxonomic group that sits below a **kingdom** and above a **class**. See page 74.

Phytoplankton (Gr. "little wandering plants"): unicellular marine algae that carry out perhaps one third to one half of Earth's **photosynthesis**. Many are eukaryotic and closely related to the green and brown seaweeds, but some are bacterial.

Plurilocular (L. "many small places"): a many-celled **sporangium** in brown seaweeds in which reproductive cells develop by cell replication (**mitosis**).

Polymer (Gr. "many parts"): a molecule made of repeating units, like links in a chain.

Polysiphonous (Gr. "many pipes"): a seaweed body plan in which each **axis** is several cells thick.

Receptacle (L. "a container"): a visible swelling in fertile seaweeds that contains many **conceptacles**. See photographs on pages 9 and 200.

Respiration (L. "breathing"): in biochemical terms, the burning of food in oxygen to release energy, water, and carbon dioxide.

Rhizoid (Gr. "root"): a **holdfast** or cell that looks like a plant root (see page 79).

Siphonous (Gr. "a pipe or tube"): in green seaweeds, used to describe single cells that contain many nuclei. Not to be confused with **monosiphonous** or **polysiphonous**, which are terms used to describe red seaweeds (see page 78).

Sorus (Gr. "a heap"): a patch of **sporangia**.

Speciation (L. "turning into forms"): the umbrella term for the various processes through which new **species** are formed.

Species (L. "a form"): the smallest taxonomic group, sitting below a **genus**. Members of a species can mate with other members to reproduce. See p. 74.

Spermatangium (Gr. "a seed container"): a red seaweed structure that makes their male **gametes** (**spermatia**). It is the male **gametangium** in red seaweeds.

Spermatia (Gr. "seed"): the sperm of red seaweeds. They are not called "sperm" because they lack tails (**flagella**) and cannot move much.

Spore (Gr. "seed"): a reproductive cell that never combines with another reproductive cell but nonetheless grows into a multicellular generation (as opposed to a **gamete**).

Sporophyll (Gr. "the leaf that makes spores"): a specialized **blade** that makes reproductive structures.

Sporophyte (Gr. "the plant that makes spores"): the life cycle generation that produces **spores**, usually in its **sporangia**.

Stipe (L. "a post or tree trunk"): the stalk of a seaweed.

Stolon (L. "a shoot"): a horizontal strand that supports multiple upright **fronds** spaced at regular intervals (see page 79).

Stramenopile (L. "straw-like hairs"): a major eukaryotic group that is part of the kingdom Chromista (see p. 74).

Symbiont, symbiosis (Gr. "living together"): an organism that lives in intimate association with another, where both organisms benefit from living together.

Sympodial growth (Gr. "shared foot"): a type of growth in which the main branch of the seaweed stops growing once a branch appears. The branch then takes over the main growing duties, handing them on to one of its branches in turn.

Taxonomy (Gr. "the law of arrangement"): the study of classification. Taxonomists often work out **phylogenies** to help understand where organisms have come from.

Tetraspore (Gr. "a cluster of four seeds"): the asexual reproductive cells that are released by the **tetrasporophyte**. Once released, they grow into the **gametophyte** generation.

Tetrasporophyte (Gr. "the plant that bears the tetraspores"): a **diploid** multicellular life cycle stage in the red class Florideophyceae that develops from the **carpospores** released by the **carposporophyte** and which, in turn, produces **tetraspores**.

Thallus (Gr. "shoot"): the main body of any seaweed. In botany, thallus is tissue that doesn't have true vascular tissue.

Transition zone: the region where the **stipe** changes into the **frond**.

Trichogyne (Gr. "a female hair"): in red seaweeds, a hair that emerges from the **carpogonium** to catch **spermatia**.

Trichothallic (Gr. "hairy shoot"): a type of growth in which a small hair emerges from a cell and both the hair and cell then expand by **intercalary** growth at the base of the hair (see page 79).

Uniaxial (L. "one direction"): a seaweed whose body is made up of one main strand, which may or may not have branches (see page 78).

Unilocular (L. "one small place"): a single-celled **sporangium** in brown seaweeds in which reproductive cells develop by cell division (**meiosis**).

Utricle (L. "a small, bottle-shaped sac"): a bottle-shaped blob that pinches out from a filament in some green seaweeds. The utricles stack together, like bottles, to create the surface of the seaweed (see photograph on page 85).

Zoöspore (Gr. "a mobile seed"): an older term for a **meiospore**.

Zygote (Gr. "yoked together"): the single **diploid** cell that results from the fusion of **haploid** male and female **gametes**.

Zygotospores (Gr. "the seeds that come from fusion"): the **diploid** reproductive cells that are released by the fertilized **carpogonium** in some members of the order Bangiales and grow into the **conchocelis**.

RESOURCES

BOOKS (GENERAL)

Braune, Wolfgang, translated and revised
by Guiry, Michael D. 2011
Seaweeds.
Koeltz https://koeltz.com

Freese, Jillian. 2018
A is for Algae.
Jillian Freese Publications
A book for children

Gee, Henry. 2021
A (Very) Short History of life on Earth.
Picador.
For more background on deep time.

Iselin, Josie. 2019
The Curious World of Seaweed.
Heyday.

Pereira, Leonel. 2016
Edible Seaweeds of the World
CRC Press

Shetterly, Susan H. 2018
Seaweed chronicles: a world at the water's edge.
Algonquin

BOOKS (ACADEMIC)

Graham, Linda E., James M. Graham,
& Lee Warren Wilcox. 2009
Algae (2nd edition). Benjamin Cummings.
This is an undergraduate-level textbook.

FIELD GUIDES

ANTARCTICA
Schories, Dirk. 2016
Marine Wildlife King George Island Antarctica.
Dirk Schories Publications

ATLANTIC
Bunker, Francis StP. D., Juliet A. Brodie,
Christine A. Maggs, & Anne R. Bunker. 2010
Seasearch guide to seaweeds of Britain and Ireland.
Marine Conservation Society.

Zwamborn, Miek, translated by Michele Hutchison. 2020
The Seaweed Collector's Handbook: From Purple Laver to Peacock's Tail.
Profile Books.

INDIAN OCEAN
Richmond, Matthew (ed). 1997
A Guide to the Seashores of Eastern Africa

and the Western Indian Ocean Islands.
SIDA/Department for Research Cooperation, SAREC.

PACIFIC
Druehl, Louis. 2001
Pacific Seaweeds.

Harbour
Nelson, Wendy. 2013
New Zealand Seaweeds: an illustrated guide.
Te Papa Press

PHYCOLOGICAL SOCIETIES

A number of phycological societies exist worldwide.
These are primarily academic.

Phycological Society of America (PSA)
www.psaalgae.org

British Phycological Society (BPS)
brphycsoc.org

Japanese Society of Phycology
sourui.org/JSPEnglish/welcome.html

Korean Society of Phycology
www.algae.or.kr/

USEFUL ONLINE SOURCES

Algaebase; the gold standard academic site for algal taxonomy
www.algaebase.org/

Algae-L; a bulletin board to ask questions about algae
www.seaweed.ie/algae-l/

FAO (Food and Agriculture Organization of the United Nations)
www.fao.org (and use the search bar)
For information about mariculture

OBIS (Ocean Biodiversity Information System)
obis.org
For information about species distribution

Timetree: the timescale of life
www.timetree.org/
For information about evolutionary distances

The Seaweed Site
www.seaweed.ie

INDEX

ACKNOWLEDGMENTS

This book is very much a collaborative effort. In a literary capacity, Anna Southgate has edited and improved the text throughout. Jane Smith collated the photos, John Woodcock drew the diagrams, and Ginny Zeal designed the layout. Jacqui Sayers and Sophie Collins helped to organize things at the start and Emily Angus at Quarto steered the whole process.

Professionally, Susan Brawley (Maine, USA) read an early draft and her comments significantly improved the book. My thanks go to Julian Griffin and Ned Mason (Aberdeen, UK) and Giovanni Furnari (Catania, Italy) for commenting on certain sections. Christine Maggs (JNCC, UK), for whom *Plocamium maggsiae* is named, and Mike Guiry (Galway, Ireland), for whom *Fucus (macro)guiryii* is named, will always be phycological inspirations.

Academically, I'd like to thank and apologize to my research group, Akanksha Agrawal, Jessie Cuomo, Alexander Goodridge, and Marie Rapin, for putting up with slow replies while I was writing this.

Finally, particular thanks go to Victoria James for telling me to write a book, Katharine Sanderson for showing me how to, and Kathryn Stevens for buying me something to write this one in.

If I could dedicate the whole thing it would—as always—be to my family.

PICTURE CREDITS

The publisher would like to thank the following for permission to reproduce copyright material. All reasonable efforts have been made to trace copyright holders and to obtain their permission for the use of copyright material. The publisher apologizes for any errors or omissions and will gratefully incorporate any corrections in future reprints if notified.

l=left; r=right; t=top; b=bottom; m=middle; i=inset.

2 F1 Online/Shutterstock; 6–7, 201t Lou Luddington/imagequestmarine.com; 8l Encyclopaedia Britannica/UIG Via Getty Images; 8r HERVE CONGE, ISM/SCIENCE PHOTO LIBRARY; 9t, 15 BIOPHOTO ASSOCIATES/SCIENCE PHOTO LIBRARY; 9b Gary Bell/Oceanwide/naturepl.com; 11 Old Books Images/Alamy Stock Photo; 12 StockBURIN/Shutterstock; 13i, 31i GERD GUENTHER/SCIENCE PHOTO LIBRARY; 13b, 137, 143i, 168, 169t, 170t, 172, 190, 191, 193, 201b agefotostock/Alamy Stock Photo; 17tl N. J. Butterfield, Department of Earth Sciences, University of Cambridge; 17tr Photo courtesy of Dr Qing Tang, working at Virginia Tech, USA; 17b Zoonar GmbH/Alamy Stock Photo; 19 WILLEM KOLVOORT/NATURE PICTURE LIBRARY/SCIENCE PHOTO LIBRARY; 20 Phil McDermott/Alamy Stock Photo; 23 Victor Harris/Alamy Stock Photo; 24, 27l The Natural History Museum/Alamy Stock Photo; 25 roger tillberg/Alamy Stock Photo; 26l Florilegius/Alamy Stock Photo; 26i Zuri Swimmer/Alamy Stock Photo; 27i IanDagnall Computing/Alamy Stock Photo; 28 NANO CREATIVE/SCIENCE SOURCE/SCIENCE PHOTO LIBRARY; 29l, 38m DENNIS KUNKEL MICROSCOPY/SCIENCE PHOTO LIBRARY; 29i NATIONAL CANCER INSTITUTE/SCIENCE PHOTO LIBRARY; 31l Science History Images/Alamy Stock Photo; 32 DeAgostini/Getty Images; 38l, 87, 129i, 142, 143t, 145, 152, 153, 154–155, 157i, 163i, 177t SR_13_John Huisman/Algaebase; 38r X-ray image courtesy of Prof. Egbert Boekema, Groningen, The Netherlands; 41 Stocktrek Images, Inc./Alamy Stock Photo; 42 Tui De Roy/naturepl.com; 43l, 83 Amar and Isabelle Guillen - Guillen Photo LLC/Alamy Stock Photo; 43i ALDERNEYMAN/Alamy Stock Photo; 43r, 53t, 114tr, 126, 150t, 161t, 192 Premaphotos/Alamy Stock Photo; 45, 49l, 224, 230–231 BIOSPHOTO/Alamy Stock Photo; 47, 139, 173t, 181i, 231i SR_014_Colin Bates/Algaebase; 49r 22August/Shutterstock; 51t; Brent Stephenson/naturepl.com; 51b, 63, 133, 134 Hemis/Alamy Stock Photo; 53b SR_028_Frederic Leliaert/Algaebase; 54 Edmund Lowe Photography/Getty Images; 56 Chuck Painter/Stanford News Service; 57l Les Gibbon/Alamy Stock Photo; 57r John Ewing/Portland Press Herald via Getty Images; 59 Patrick Aventurier/Getty Images; 60l MAXINE ADCOCK/SCIENCE PHOTO LIBRARY; 60r, 205b Philippe Clement/naturepl.com; 61t Stockbyte/ Getty Images; 61b MATTHEW OLDFIELD/SCIENCE PHOTO LIBRARY; 62 bamboovn/ Shutterstock; 64l Sasimoto/Alamy Stock Photo; 64r Johnfoto18/Shutterstock; 65t Newscom/Alamy Live News; 65b Pedro arquero/Getty Images; 66 MATTEIS/LOOK AT SCIENCES/SCIENCE PHOTO LIBRARY; 67 Jim West/imageBROKER/imagequestmarine.com; 68l Aerial Archives/Alamy Stock Photo; 68i Brett Monroe Garner/Getty Images; 69 Auscape/Universal Images Group via Getty Images; 70l Georg Stelzner/imageBROKER/imagequestmarine.com; 70i Jean-Pierre REY/Gamma-Rapho via Getty Images; 71t, 100, 148, 199 Nick Upton/naturepl.com; 71b, 94, 95, 102, 104, 106i, 110, 112, 113, 114i, 116, 118, 121i, 124i, 124b, 125, 127t, 128, 135, 140, 146, 147i, 156, 157t, 160, 161i, 162, 173i, 178, 179i, 183i, 194,

210, 211i, 221i, 225, 228r SR_002_Mike Guiry/Algaebase; 73 Alexis Rosenfeld/Getty Images; 76–77, 188, 209i Alex Mustard/naturepl.com; 82 Techa Tungateja/Alamy Stock Photo; 84 Arvydas Kizevicius/Alamy Stock Photo; 85t Poelzer Wolfgang/Alamy Stock Photo; 85i SR_021_David Fenwick/APHOTOMARINE; 86 ANDREW J. MARTINEZ/SCIENCE PHOTO LIBRARY; 88 scubaluna/ Shutterstock; 89 WaterFrame/Alamy Stock Photo; 90 ALEXANDER SEMENOV/SCIENCE PHOTO LIBRARY; 91 Marli Wakeling/Blue Planet Archive/imagequestmarine.com; 92, 105t, 119b, 121l Arterra Picture Library/Alamy Stock Photo; 93t ARCTIC IMAGES/Alamy Stock Photo; 93i buccaneer/Alamy Stock Photo; 96l, 101t Sue Daly/naturepl.com; 96i Chris Carter; 97, 180–181, 226 Minden Pictures/Alamy Stock Photo; 101b Trustees of the Natural History Museum, London; 103t StockFood/Baranowski, Andre; 103b Heritage Image Partnership Ltd/Alamy Stock Photo; 105i, 115, 120, 127i, 136, 138, 141, 147t, 163t, 170i, 176, 177b, 183l, 189 SR_011_Ignacio Bárbara/Algaebase; 106t Jennifer Booher/Alamy Stock Photo; 107 Tim Gainey/Alamy Stock Photo; 108–109 Lee Rentz/Alamy Stock Photo; 111 Dorling Kindersley Ltd/Alamy Stock Photo; 117i SR_021_David Fenwick/APHOTOMARINE/Algaebase; 117b, 149, 166 Seaphotoart/Alamy Stock Photo; 119i Richard Wayman/Alamy Stock Photo; 123 Darwin Plus grant DPLUS068 Building foundations to monitor and conserve Falklands marine forest habitats awarded to Juliet Brodie, Natural History Museum, London; 129t, 215 David Fleetham/Alamy Stock Photo; 130, 179b Sabena Jane Blackbird/Alamy Stock Photo; 131 Dariusz Jarzabek/Shutterstock; 132 SR_022_Leila Hayashi/Algaebase; 150i WIM VAN EGMOND/SCIENCE PHOTO LIBRARY; 151 SR_021_David Fenwick/APHOTOMARINE; 159t, 159b, 174, 175 SR_016_Dirk Schories/Direct; 167t Laura Dts/Shutterstock; 167i Sergio Hanquet/naturepl.com; 169b Robert Harding/Douglas Peebles; 171 Minden Images/D.P. Wilson; 182 Nick Upton/Alamy Stock Photo; 184 Image navi - QxQ images/Alamy Stock Photo; 185t The Asahi Shimbun via Getty Images; 185i SR_018_Keoki and Yuko Stender/Marine Life Photography.com/Direct; 186 Solvin Zankl/naturepl.com; 187t F-Stop boy/Alamy Stock Photo; 187i RODRIGO ARANGUA/AFP via Getty Images; 195 Richard Robinson/naturepl.com; 196–197 Piter Lenk/Alamy Stock Photo; 197i Chris Mattison/Alamy Stock Photo; 198 Naomi Roe/Alamy Stock Photo; 200 PHILIP SMITH/Alamy Stock Photo; 202t mark Phillips/Alamy Stock Photo; 202i, 205t Photimageon/Alamy Stock Photo; 203 Andrey Nekrasov/Alamy Stock Photo; 204 Science Photo Library/Alamy Stock Photo; 207t DR JEREMY BURGESS/SCIENCE PHOTO LIBRARY; 207i Steve Taylor ARPS/Alamy Stock Photo; 207b Manfred Gottschalk/Alamy Stock Photo; 209 FLPA/Shutterstock; 211l Nature Production/naturepl.com; 212 DR KEITH WHEELER/SCIENCE PHOTO LIBRARY; 213 Julian Cremona/Alamy Stock Photo; 214 Franco Banfi/naturepl.com; 216 mauritius images GmbH/Alamy Stock Photo; 217 F1online digitale Bildagentur GmbH/Alamy Stock Photo; 218 MelindaChan/Getty Images; 219t UCSD SC&A - C. K. Tseng in a diving suit at Laguna Beach, California./Schaffer, Victor B.. SIO Photographs Collection. SMC 59. Special Collections & Archives, UC San Diego/Special Collections & Archives, UC San Diego, La Jolla, 92093-0175 (https://lib.ucsd.edu/sca); 219i Irina Kryvasheina/Alamy Stock Photo; 221t Malcolm Schuyl/Flpa/imageBROKER/Shutterstock; 223 Blue Planet Archive/Alamy Stock Photo; 227 Minden Images/Yasuyuki Koike; 228l Nature in Stock/Alamy Stock Photo; 229 Daniel Poloha Underwater/Alamy Stock Photo.